Spacecraft System Health Management: Current Practices and Emerging Technologies

Spacecraft System Health Management: Current Practices and Emerging Technologies

By Samir Khan

400 Commonwealth Drive
Warrendale, PA 15096-0001 USA
E-mail: CustomerService@sae.org
Phone: 877-606-7323 (inside USA and Canada)
           724-776-4970 (outside USA)
Fax: 724-776-0790

Library of Congress Catalog Number 2024945945
http://dx.doi.org/10.4271/9781468607802

ISBN-Print   978-1-4686-0779-6
ISBN-PDF   978-1-4686-0780-2
ISBN-epub   978-1-4686-0781-9

To purchase bulk quantities, please contact: SAE Customer Service

E-mail: CustomerService@sae.org
Phone: 877-606-7323 (inside USA and Canada)
            724-776-4970 (outside USA)
Fax: 724-776-0790

Visit the SAE International Bookstore at books.sae.org

Publisher
Sherry Dickinson Nigam

Product Manager
Amanda Zeidan

Production and
Manufacturing Associate
Michelle Silberman

To my family, both near and far, whose love and encouragement have shaped my journey in countless ways

Contents

A Note from the Author!

Ensuring the health and reliability of spacecraft systems is not just a technical challenge; it is a fundamental necessity for mission success and the safety of human life in space. This concept has been vividly captured in popular culture, such as Stanley Kubrick's "2001: A Space Odyssey," which left a mark on our imagination by presenting a future where technology transforms every aspect of human existence. The film is renowned for its portrayal and the questions it raises about humanity's path; it highlights our increasing reliance on technology and our interest in the unknown. Among its most memorable aspects is the autonomous onboard computer—HAL, whose ability to understand and predict the behavior of spacecraft equipment underscores a future where computers play an important role in ensuring human safety.

Fast forward half a century, and we find ourselves in need of these types of systems. The tragedy of the Space Shuttle accident, which resulted in the loss of 14 astronauts, emphasizes the urgent need for advanced monitoring and protection systems like HAL. This need has triggered the development of technologies that were once part of fiction. Today, integrating human–computer interaction has become a reality, thanks to a suite of technologies like OpenAI, Google DeepMind, and IBM Watson, designed to meet our complex needs.

Many of these advancements are particularly relevant to spacecraft, where the failure of equipment is a significant challenge.

Despite rigorous testing, spacecraft components—often manufactured by different companies—can fail shortly after deployment. This issue of "infant mortality" has puzzled engineers for decades and has led to substantial financial losses. Surprisingly, the space industry has traditionally accepted these failures as part of the business, with one in four satellite owners filing major insurance claims within the first year of operation. This acceptance has financial implications, costing taxpayers over $10 billion annually due to failure.

However, the emergence of system health management technologies offers some hope. Engineers, drawing on their expertise in equipment design and performance, are now making connections between early failure signs and the actual breakdowns, uncovering manufacturing and testing flaws. This insight is a game changer, offering a way to detect and prevent these failures and save significant costs. Unlike the consumer electronics industry, where warranties and returns offer consumers protection against defects, the spacecraft industry lacks such safeguards. The gap places a heavy burden on agencies and operators who must rely on insurance to cover the costs of failures. Yet, introducing system health management promises to shift this paradigm, offering a layer of protection.

This book takes a unique approach to discussing this topic. Moving away from the dense technical jargon and mathematical models typically associated with this field, it opts for a more accessible, discussion-oriented style. This method aims to unravel the complexities surrounding spacecraft failures, making the subject more engaging and understandable. By focusing on the broader implications and practical aspects of system health management, this book invites readers to explore the nuances of a problem that has long challenged the space industry, offering fresh insights into a critical area of space exploration.

To summarize some important points from the book:

- The evolution of space technology: The transition from government-led to private sector-driven space exploration has democratized access to space, fostering innovation and reducing costs. This shift underscores the need for robust monitoring systems that can adapt to the increasing complexity and diversity of missions.

- Collaboration between public and private sectors: The burgeoning collaboration between government agencies and private companies is fostering rapid innovation and expanding a number of possibilities in space activities.

- Complexity of spacecraft systems: Understanding the subsystems is crucial for effective health management. As spacecraft become more sophisticated, integrating health management systems becomes necessary for mission success.

- Importance of proactive health management: The ability to predict and prevent failures before they occur can significantly enhance mission safety and reliability. This proactive approach is facilitated by advanced diagnostics, telemetry data analysis, and the application of artificial intelligence (AI) and machine learning (ML).

- Advancements in fault management and data analysis: Techniques such as fault tree analysis (FTA), model-based reasoning, and data-driven diagnostics are all still essential for identifying and mitigating potential failures.

- The role of AI: The application of AI in spacecraft health management is transformative. Such technologies enable the analysis of vast amounts of data, offering unprecedented insights into system health and potential failures.

- Future technologies: The exploration of digital twins (DTs) and autonomous capabilities has a lot of potential. These technologies promise to revolutionize the way spacecraft are designed, monitored, and maintained.

Foreword

Houston, we have a problem.

On April 14, 1970, Apollo 13 sent this urgent message to NASA's Mission Control Center in Houston, sparking the start of a dramatic rescue mission that captivated the world. Mission Control faced immense challenges in identifying and rectifying the failure on the spacecraft. The only available information came from various sensor readings displayed in the cockpit and the crew's health status. Despite limited resources, Mission Control used a wide range of tools, both on the ground and inside the spacecraft, to replace the shut-down computer and conserve power, ultimately staging a remarkable recovery that brought the crew safely back to Earth. This survival story, often referred to as "a Successful Failure," remains one of the most interesting episodes in the Apollo program.

The history of space activities has been a relentless struggle to maintain system integrity and achieve predetermined missions in the extreme environment of space. Many unmanned and manned space systems operate as "non-repairable systems," where human intervention for repairs is impossible once they break down. This harsh reality rarely occurs with ground-based systems. For manned missions like Apollo 13, a major system failure or a failure to promptly restore normal operations can result in loss of life. For unmanned systems, such as artificial satellites, the loss can lead to

devastating financial and operational impacts for governments and companies that invest heavily in their development and launch. Therefore, it is crucial for spacecraft to survive and continue their missions at all costs.

This book explores various methodologies to maintain spacecraft health and increase mission success rates. It focuses on diagnosing failures ("diagnosis") and predicting potential future failures ("prognosis"), providing a comprehensive overview of recent trends, including autonomous satellites, AI, and DTs.

The only way to determine the current health status of a satellite from the ground is through telemetry data. Any error in this judgment can result in the loss of the satellite, making this task critically important. The number of telemetry data points can range from tens to thousands, depending on the satellite's complexity. While it may be straightforward to detect anomalies from a single telemetry data point, identifying abnormalities often requires comparing multiple data points. Once an anomaly is detected, diagnosing its cause involves analyzing a larger set of telemetry information. Predicting failures, where the failure has not yet occurred but is imminent, present an even greater challenge. This book covers fault diagnosis and prognosis methodologies, from basic concepts to advanced levels, based on case studies and research examples from actual missions, and introduces the latest research using AI.

Traditionally, ground personnel have carried out fault diagnosis and prognosis. However, as onboard computers on satellites become faster, satellites can now decide autonomously. This autonomy offers an advantage over ground-based decision-making, which is limited by the speed of telemetry downlinks. Managing all satellites from the ground has become increasingly challenging, especially with the rise of small satellite constellations like Starlink, which aim to have 12,000 satellites in orbit. For deep space probes, autonomous judgment is crucial due to communication time delays. In this context, "spacecraft autonomy" is becoming increasingly important, with advancements in onboard computing and AI playing a significant role.

How do failures and their symptoms manifest? Satellite operators develop a sense of "normal" telemetry by consistently monitoring satellites under typical conditions, allowing them to quickly spot anomalies. For autonomous systems, how can a computer distinguish between normal and abnormal states? The second half of this book delves into these technical issues, discussing the history and current state of computer development, spacecraft information system architecture, and software (SW) design methods. It also highlights the importance of cybersecurity, on-orbit services, and DTs in enhancing spacecraft robustness. The final chapter evaluates various technologies and presents a vision for the future, offering guidance for readers interested in conducting research in this field.

In addition to its deep technical discussions, the book explains new trends in space development, such as New Space and small satellite constellations, and provides basic knowledge of spacecraft systems. It offers easy-to-understand explanations for readers unfamiliar with space activities or spacecraft systems. The extensive list of references allows readers to explore papers and articles in greater depth. By introducing various accidents and failures from past space missions and the lessons learned, the book raises awareness of the challenges of maintaining spacecraft health.

Space is a world of challenges. Through this book, readers will appreciate the difficulties spacecraft systems face in their operating environment and learn about the strategies used by space engineers and researchers to ensure spacecraft survival. I hope this book provides a glimpse into the dramatic and intellectually stimulating battles between humans and nature in space.

Professor Shinichi Nakasuka
University of Tokyo

Acknowledgments

This book would not have been possible without the support, guidance, and encouragement of many people to whom I am deeply grateful.

First and foremost, I want to express my heartfelt gratitude to Professor Takehisa Yairi and Dr Seiji Tsutsumi for their invaluable help with the subject area and the review process. Their expertise and insights have significantly shaped this work, and their encouragement helped me navigate through challenging parts of the writing process.

I would also like to thank my colleagues at the Japan Aerospace Exploration Agency and students at the University of Tokyo, whose enthusiasm for space exploration and intelligent systems has been a constant source of inspiration. The discussions, debates, and feedback from the Khan Laboratory have enriched my understanding and fueled my curiosity.

Special thanks go to my family, whose patience and love kept me going through countless late nights and weekends spent writing. To my mother, whose belief in me made all the difference and to my wife, whose support and encouragement inspired me to start and complete this work—this journey would not have been possible without you.

Lastly, to all the pioneers and dreamers in the field of space exploration who continuously push the boundaries of what we know: this book is dedicated to your efforts and aspirations.

List of Acronyms

3D - Three-Dimensional

A2I - Artificial Intelligence for Automation

ADC - Attitude Determination and Control

ADTQC - Anomaly Detection Timing Quality Curve

AES - Advanced Encryption Standard

AFCs - Alkaline Fuel Cells

AGC - Apollo Guidance Computer

ALICE - A Large Ion Collider Experiment

ANNs - Artificial Neural Networks

AOCS - Attitude and Orbit Control System

AOS - Advanced Orbiting Systems

AR&C - Autonomous Rendezvous and Capture

ASIC - Application-Specific Integrated Circuit

ASM - Autonomous Spacecraft Management

ASW - Avionic Software

ATVs - Automated Transfer Vehicles

BDFL - Byzantine-Fault-Tolerance Decentralized Federated Learning

BIPS - Billion Instructions per Second

BIST - Built-In Self-Test

BIT - Built-In Testing

BSC - Boundary Scan Cell

CBM - Condition-Based Maintenance

CCP - Commercial Crew Program

CCSDS - Consultative Committee for Space Data Systems

CDF - Cumulative Distribution Function

CDH - Command and Data Handling

CERN European - Organization for Nuclear Research

CLSP - Commercial Lunar Payload Services

CND - Cannot Duplicate

CNES - Centre National d'Etudes Spatiales

CNNs - Convolutional Neural Networks

COMS - Communications Subsystem

COSTAR - Corrective Optics Space Telescope Axial Replacement

COTS - Commercial Off-the-Shelf

COTS - Commercial Orbital Transportation Services

CPU - Central Processing Unit

CSR - Crew Safety and Reliability

DARPA - Defense Advanced Research Projects Agency

DART - Demonstration of Autonomous Rendezvous Technology

DBNs - Dynamic Bayesian Networks

DCLOP - Deep Clustering- Based Local Outlier Probabilities

DDDS - Dynamic Data-Driven Simulation

DH - Data Handling

DL - Deep Learning

DLR - Deutsches Zentrum für Luft- und Raumfahrt

DNNs - Deep Neural Networks

DPP - Dextre Pointing Package

DSN - Deep Space Network

DSPs - Digital Signal Processors

DT - Digital Twin

ECLSS - Environmental Control and Life Support System

ECSS - European Cooperation for Space Standardization

EDAC - Error Detection and Correction

EDL - Entry, Descent, and Landing

EKF - Extended Kalman Filter

EO - Earth Observation

EOL - End-of-Life

ESA - European Space Agency

ESA-AD - ESA Anomalies Dataset

ESA-ADB - European Space Agency Anomaly Detection Benchmark

ETA - Event Tree Analysis

EVAs - Extravehicular Activities

EWMA - Exponentially Weighted Moving Average

FCM - Fuzzy C-Means

FCRs - Fault Containment Regions

FDIR - Fault Detection, Isolation, and Recovery

FEA - Finite Element Analysis

FMEA - Failure Mode and Effects Analysis

FPGA - Field-Programmable Gate Array

FTA - Fault Tree Analysis

FTP - Fuel Turbopump

GA - Genetic Algorithms

GASPACS - Get Away Special Passive Attitude Control Satellite

GDC - Gemini Digital Computer

GEO - Geostationary Earth Orbit

GNC - Guidance, Navigation, and Control

GNSS - Global Navigation Satellite System

GPR - Gaussian Process Regression

GPUs - Graphical Processing Units

HA - Hazard Analysis

HAZOP - Hazard and Operability

HGA - High-Gain Antennas

HI - Health Indicators

HMI - Human-Machine Interface

HMMs - Hidden Markov Models

HMS - Health Monitoring Systems

HNNs - Hopfield Neural Networks

HPTM - High Priority Telemetry

HSMs - Hardware Security Modules

HST - Hubble Space Telescope

HUDS - Heads-Up displays

HUMS - Health and Usage Monitoring Systems

HW - Hardware

IC - Integrated Circuit

ICU - Instrument Control Computer

IDS - Intrusion Detection Systems

IEEE - Institute of Electrical and Electronics Engineers

IET - Institution of Engineering and Technology

I/O - Input/Output

IoT - Internet of Things

iRODS - Integrated Rule-Oriented Data System

ISA - Instruction Set Architecture

ISS - International Space Station

JAXA - Japan Aerospace Exploration Agency

JSPS - Japan Society for the Promotion of Science

JWST - James Webb Space Telescope

KF - Kalman Filter

kNNs - k-Nearest Neighbors

LCL - Latching Current Limiter

LEO - Low Earth Orbit

LM - Lunar Module

LOC - Loss of Crew

LOM - Loss of Mission

LRO - Lunar Reconnaissance Orbiter

LSTM - Long Short-Term Memory

MA - Moving Average

MBSE - Model-Based System Engineering

M-Code - Military Code

MIPS - Million Instructions per Second

ML - Machine Learning

MLPs - Multi-Layer Perceptrons

MMS - Multi-Mission Modular Spacecraft

MRO - Mars Reconnaissance Orbiter

MSL - Mars Science Laboratory

MTTR - Mean-Time-to-Repair

NARX - Nonlinear Autoregressive Networks with Exogenous Inputs

NASA - National Aeronautics and Space Administration

NFF - No Failure Found

OBCs - Onboard Computers

OBSW - Onboard Software

OLS - Ordinary Least Squares

OMV - Orbital Maneuvering Vehicle

OOL - Out-of-Limits

OPS-SAT - Operations Satellite

ORI - Orbital Replacement Instrument

ORU - Orbital Replacement Unit

OS - Operating System

OSA - Open System Architecture

OSA-CBM - Open System Architecture for Condition-Based Maintenance

OSS - Orbital Satellite Services

PCA - Principal Component Analysis

PCDU - Power Control and Distribution Unit

PCU - Power Control Unit

PDA - Probabilistic Design Analysis

PDF - Probability Density Function

PEM - Proton Exchange Membrane

PF - Particle Filter

PHM - Prognostics and Health Management

PLSDA - Partial Least Squares Discriminant Analysis

PRA - Probabilistic Risk Assessment

PROBA - Project for Onboard Autonomy

PSO - Particle Swarm Optimization

R2 - Robonaut 2

RBAC - Role-Based Access Control

RBDs - Reliability Block Diagrams

RBFNN-AEKF - Radial Basis Function Neural Network-Aided Adaptive Extended Kalman Filter

RHBD - Radiation-Hardened by Design

RHBP - Radiation-Hardened by Process

RISC - Reduced Instruction Set Computer

RMS - Remote Manipulator System

RNN - Recurrent Neural Networks

RRM - Robotic Refueling Mission

RTGs - Radioisotope Thermoelectric Generators

RTOS - Real-Time Operating Systems

RTPF - Real-Time Particle Filters

RUL - Remaining Useful Life

SA - Simulated Annealing

SAR - Synthetic Aperture Radar

SARA - System Analysis and Risk Assessments

SCV - Spacecraft Control Vector

SDRs - Software-Defined Radios

SDSs - Software-Defined Satellites

SDT - Spacecraft Digital Twin

SEEs - Single-Event Effects

SEFIs - Single Event Functional Interrupts

SELs - Single Event Latchups

SETs - Single Event Transients

SEUs - Single-Event Upsets

SHM - System Health Management

SIDs - Structure IDs

SIFA - System Integration Failure Analysis

SLS - Space Launch Systems

SMC - Sequential Monte Carlo

SMM - Solar Maximum Mission

SMs - Servicing Missions

SoC - Systems-on-Chip

SOEs - Spacecraft Operations Engineers

SOFC - Solid Oxide Fuel Cell

SPDM - Special Purpose Dexterous Manipulator

SRAM - Static Random-Access Memory

SRM - Solid Morton-Thiokol Rocket Motor

SSH - Secure Shell

SSRMS - Space Station Remote Manipulator System

SVF - Scenario Verification Facilities

SVMs - Support Vector Machines

SW - Software

SysML - Systems Modeling Language

TID - Total Ionizing Dose

TMR - Triple Modular Redundancy

TMTC - Telemetry and Telecommand

TPS - Thermal Protection System

TPU - Tensor Processing Unit

TRON - Testbed for Rendezvous and Optical Navigation

TVC - Thrust Vector Control

UKF - Unscented Kalman Filter

UQ - Uncertainty Quantification

US - United States

USAF - United States Air Force

UT - Unscented Transform

VIO - Visual-Inertial Odometry

VPU - Vision Processing Unit

WFPC2 - Wide Field and Planetary Camera 2

XAI - Explainable Artificial Intelligence

Any sufficiently advanced technology is
indistinguishable from magic.

—Arthur C. Clark

Over the last two decades, the landscape of space exploration and satellite technology has undergone a significant transformation, primarily fueled by the global challenges it seeks to address. This transition is not just techno-logical. Today, we focus on creating reliable, sustainable, and cost-effective systems. This requires innovations capable of facilitating long-distance communication, ensuring prolonged operational service, and integrating intelligence into spacecraft management systems [1.1]. In the 1600s, the invention of the telescope transformed our understanding of the universe and marked the start of modern space technology. The field has grown since then, especially with the first space probe launched in 1957. This event paved the way for global collaboration and advance-ments in various areas, including medicine, communication, transportation, and environ-mental studies. Satellites have been particularly transformative. They provide us with vital infor-mation about our planet's environment, weather, and climate change. They are additionally vital in today's communication and navigation systems. The Hubble Space Telescope (HST), for example, has given us detailed views of the universe, deepening our knowledge of its laws. Now, with the growing need for fast Internet, compre-hensive Earth observation (EO), and worldwide connectivity, satellite technology innovation is more important than ever. This demand has led to the creation of various types of satellites, like those for communication, observation, navigation, and remote sensing. Many private companies, such as Boeing, O3b, Telesat, ViaSat, SpaceX, OneWeb, Audacy, Karousel, Kepler, and Theia, are now leading in space exploration and commercialization. There is even talk of commercial travel for everyday people, showing a major shift in industry.

With the advancement of technology, space missions are becoming more complex, with spacecraft traveling further and faster than ever. This poses engineering challenges, especially to achieve high stability and reliability [1.2]. Despite detailed preparation and design, mission failures can occur, often exacerbated by unpredictable and harsh operating environments. These mistakes not only jeopardize mission success but also result in financial losses or tragic outcomes, as seen in the Challenger explosion in 1986 and the Columbia disaster in 2003. To manage these risks, it is important to maintain system reliability. Traditionally, this involved isolating issues quickly after detection and designing systems that are fault tolerant. However, with increasing design complexities and heightened requirements, predicting and understanding fault propagation is becoming more and more challenging. This is where predictive technology comes into play.

System health management (SHM), initially demonstrated in the aerospace industry, is crucial for ensuring the reliability and safety of complex systems. This approach involves setting up a wide network of sensors across a system to collect detailed data on its condition. By enabling the interaction and analysis of these data, it is possible to detect and address potential failures before they happen. As technology progresses, there is an increasing push toward making these monitoring and diagnostic processes automatic. This trend extends beyond space exploration and is also becoming vital in other critical areas, such as submarine operations and environments with high radiation levels.

In space missions, spacecraft health is necessary for the success and safety of the journey. It signifies the spacecraft's ability to function as intended, with all systems, subsystems, and components performing their designated roles. Spacecraft health management is a sophisticated activity that extends beyond mere monitoring. It provides the spacecraft with the capability to assess its health, diagnose issues independently, and, in case of problems, take corrective actions autonomously. This could entail fixing the issue to return to normal operations or taking measures to reduce risks to the mission and crew. The core of this management approach is continuous monitoring and diagnosis. It serves as the foundation for informed decision-making and executing necessary actions based on the spacecraft's current state.

Merging SHM with automated capabilities in spacecraft operations represents a significant advancement. It not only strengthens mission safety and success amid growing complexities, but also exemplifies the potential of applying such technologies across various high-stakes domains. This seamless fusion of monitoring, diagnosis, and independent correction underscores a proactive strategy to maintain system integrity and functionality, ensuring that missions can proceed safely and efficiently.

This book explores the benefits of automating these processes. It investigates various technical concepts essential for successful implementation and highlights the methodologies impacting its application. To provide a comprehensive overview of current technological advancements and business models in the field, we also aim to explore the future of spacecraft maintenance technology. Such technology holds significant economic implications. Not only is the development and implementation of such technology responsible for creating many jobs, but it also plays a significant role in the revenue generated from spacecraft launches. The aim is to help readers grasp the impact of these technologies on the space industry, technologically and economically. By concentrating on the latest developments and prospects, the book provides valuable insights for professionals, researchers, and enthusiasts interested in the evolving landscape of spacecraft maintenance and health management. It emphasizes the significance of these innovations in enhancing the safety, efficiency, and reliability of space missions, while also highlighting their role in driving economic growth in the aerospace sector.

1.1.
The Shifting Landscape

Space exploration, once dominated by national governments because of their resources and expertise, has now seen a major shift with the emergence of private companies such as SpaceX, Blue Origin, and Virgin Galactic. They have taken advantage of technological advancements to construct rockets and spacecraft more cost-effectively than traditional government programs. The change resulted in more private space enterprises, each seeking a larger market share. The impact of this shift is substantial. For instance, Blue Origin's New Shepard rocket, aimed at suborbital tourism flights, is poised to open a new market in space travel. Similarly, SpaceX's Falcon 9 rocket, designed for reusability, has significantly reduced launch costs, showcasing the capability of nongovernmental companies to manage space missions as effectively as government agencies. This success can be partly attributed to the private sector's

agility in responding to customer needs and carving out niche markets for their services [1.3, 1.4]. In contrast, government space agencies like the National Aeronautics and Space Administration (NASA) often face bureaucratic hurdles and longer timelines, potentially slowing down innovation. However, the increased competition from more players in the field could lead to faster innovation and more rapid technological advancements, benefiting the entire realm of space exploration (Figures 1.1 and 1.2).

An engaging development is the increasing inclination to adopt smaller, more agile spacecraft from monitoring the Earth's environment to offering Internet connectivity to remote areas. The current shift toward "smaller" assets also has implications for the industry. Smaller spacecraft are more flexible and can be deployed more quickly. This shift toward "smaller" assets carries significant implications for the industry. A prime example is the CubeSat, a type of miniaturized spacecraft. CubeSats, usually less than 10 cm in size and weighing less than 1.33 kg, are increasingly used by academic institutions and small companies to experiment with new space technologies. Their compact size makes it easier and cheaper to design, build, and launch, significantly lowering the barrier to entry in the space sector. But this trend is not just about developing smaller satellites. As a result, we are likely to see a surge in space-based projects, ranging from scientific research to commercial applications, driven by this new era of compact and cost-effective spacecraft.

The entry of private companies into the space industry, with a focus on building smaller spacecraft, reflects the democratization of space access. Historically, only a few countries had the resources and expertise for space missions. Today, however, we observe an increasing

Figure 1.1 The International Space Station (ISS) was assembled in space over 10 years and 30 missions.

NASA

Figure 1.2 A preview for future space travel. The SpaceX Dragon 2 is the only human-rated orbital transport spacecraft whose primary role is to ferry crews to the ISS.

NASA

number of countries, and even private individuals, capable of reaching space and a broader spectrum of participants following their unique interests. For example, a small company might prioritize launching a satellite constellation for weather monitoring over government space agencies.

While this democratization opens various opportunities, it additionally brings forth concerns about safety and regulation. Government space agencies usually follow strict safety standards. However, private organizations may not face the same level of scrutiny, which could raise the risk of accidents and casualties. The trend toward smaller spacecraft raises questions about their capability to perform complex, long-duration missions, considering their limited resources and payload capacity. The growing accessibility of space gives rise to concerns about space debris and the potential for collisions. As more agencies launch satellites, the risk of space congestion and debris generation increases, posing a threat to all space operations. The possibility of space being used for military by some nations or businesses also starts a new space race, which might not align with the broader public interest.

This change from "government-led" to "private," from "large" to "small" spacecraft, and from a "few" players to "many" in the space industry brings significant implications that require careful monitoring. As the space sector continues to evolve, addressing these challenges and opportunities is crucial for its sustainable growth. However, it is undeniable that private companies have shaken up the traditional model. They have introduced quicker innovation and more frequent launches, leading to a wider variety of missions and greater diversity in the industry.

1.1.1.
A New Era in Space Technology

The United States (US) continues to lead in utilizing space technology for both defense and civil applications—this dual use has always been a major driver in the country's space program. Military applications have been in practice for decades, with the development of reconnaissance and early warning systems. In recent years, this focus has also changed toward civil applications, such as GPS navigation and EO. These advancements have already influenced our transportation, agriculture, and disaster management systems. In Europe, the use of space technology has focused more on collaborations that meet communication requirements. Reliable Internet access in remote areas is increasingly necessary for global connectivity. Governments and businesses are working together to develop cost-effective communication technologies that can be deployed rapidly. This has led to the creation of mega-constellations of small satellites for global Internet coverage. Therefore, the space industry is going through all this digital transformation, with an emphasis on SW-based flexible systems. This led to the development of SW-defined radios (SDRs) and SW-defined satellites (SDSs) that are more flexible than their hardware (HW)-based counterparts and can be reconfigured on the fly. It also enables for better utilization of resources, as well as increased efficiency and reliability. Finally, there is a shift toward the mass production of satellites. Advances in manufacturing technology, like three-dimensional (3D) printing and commercial off-the-shelf components, have enabled this. Companies such as OneWeb, Starlink, Planet, and Spire are at the forefront of mass-producing satellites that provide a range of services such as EO, communication, and navigation (Figure 1.3).

Figure 1.3 New business by small/micro-satellite constellation: (a) SpaceX Starlink: Internet in the sky with 12,000 satellites in the near future. (b) A miniaturized spacecraft, CubeSat, often uses commercial off-the-shelf components for its electronics and structure.

SpaceX.

NASA.

(a)   (b)

The rise of these satellite constellations proved to be a significant game changer in the industry. These constellations include many small, affordable satellites that collaborate and can be launched at a fraction of the cost of traditional missions. A key advantage here is the ability to offer frequent observation and communication capabilities. Multiple satellites working together can increase the frequency of EO from space. This can aid in monitoring and addressing issues such as climate change, natural disasters, and urbanization. The growing frequency of communication also permits faster and more efficient data transfer, which may prove beneficial in the range of industries, from agriculture to logistics. Of course, this ensures the ability to handle big data. However, with the use of multiple spacecraft, it is possible to manage this issue over a larger area than with a single satellite.

It should be emphasized that the driving force behind these recent innovations is the private sector's involvement in developing state-of-the-art technologies. This is because, by prioritizing innovation and new ideas, we are not limited to traditional government bureaucracy and red tape, which includes innovations such as reusable rockets, autonomous spacecraft, and 3D printing in space. Governments are also actively buying services from these private sector companies. Programs such as Commercial Orbital Transportation Services (COTS) and Commercial Lunar Payload Services (CLPS) have fostered fruitful partnerships to offer services such as cargo delivery and lunar lander development.

An informative report was recently shared by Citigroup, where they claim that the space industry is expected to achieve a revenue of $1 trillion annually by 2040, with a 95% reduction in launch costs. This decrease in costs is expected to create more opportunities for technological expansion and innovation, resulting in the development of services such as automotive production and manufacturing. The report's predictions align with similar forecasts released by Morgan Stanley, Bank of America, and other organizations. As per Space Foundation's research, the global space economy's value has grown by 70% since 2010 and reached $424 billion in 2020. Citi's report further explains that most of the satellite sector's revenue growth will originate from production. The past decade also witnessed an increase in private investments and venture capital investments in space startups. In 2020 alone, space infrastructure secured $14.5 billion of private investment. However, these ventures often encounter high upfront capital costs and the extended timeline to realize returns on space projects. The perception of space as exclusive to billionaires is a risk. The space industry needs public acceptance for broader adoption. Although private investment lowered the cost of space access, the perception of space organizations as ego-driven pet projects of the ultra-wealthy can undermine the industry's potential.

Although human spaceflight is appealing, the 2% failure rate for crewed launches is still too high for space passenger flights when compared to commercial aviation's minuscule failure rate of approximately 0.0001%. This can also be a consequence of identifying regulatory authorities that monitor the industry's growth, as several federal and international entities approve and regulate aerospace equipment. In addition, space debris is an active threat for any future launches and may impede the expansion of opportunities across the space ecosystem. Countless artificial objects orbit Earth, with more expected, some too small to track, risking the "Kessler Syndrome"—a scenario without air resistance. Debris collides, fragments, and creates an unstoppable field, preventing satellite launches. Conventional spacecraft and satellite constellations differ significantly in their design, deployment, and health monitoring approaches.

1.1.2.
Conventional Spacecraft vs. Satellite Constellations

Conventional spacecraft and satellite constellations represent two distinct approaches in the space industry, differing significantly in their design, deployment, and health monitoring strategies. Conventional systems are typically large, singular units designed for long-term missions in geostationary Earth orbit (GEO) or high Earth orbit. These missions prioritize maximizing reliability and longevity, which is reflected in their monitoring systems. Extensive prelaunch testing, onboard monitoring of critical components, and redundant systems are employed to prevent single-point failures. Maintenance opportunities are limited and often require complex servicing missions (SMs), with end-of-life management involving deorbiting or relocation to graveyard orbits.

In contrast, satellite constellations are composed of many small, standardized satellites deployed in low Earth orbit (LEO) to achieve global coverage. These constellations emphasize rapid development cycles and cost-effective, scalable designs. Health monitoring in constellations is often decentralized, using some level of autonomous health management and onboard data processing to minimize the need for continuous communication with ground stations. When individual satellites fail, they are typically replaced rather than repaired, highlighting a shorter operational lifespan and an emphasis on frequent technology refreshes. The private sector's approach to health monitoring in constellations seems to focus more on scalability and adaptability, providing a stark contrast to the conservative, high-reliability methods used in conventional spacecraft. Table 1.1 summarizes some of these differences.

Table 1.1 Comparison of conventional spacecraft and satellite constellations across key aspects.

Aspect	Conventional spacecraft	Satellite constellations
Design	Large, singular units	Many small, standardized satellites
Deployment	Individually or in small numbers	Deployed in large batches
Orbit	GEO or high Earth orbit	LEO
Mission duration	Long-term missions	Shorter operational lifespan
Development cycle	Long, with extensive testing	Rapid, with focus on cost-effectiveness
Health monitoring	Extensive prelaunch testing, onboard monitoring, redundant systems	Decentralized, autonomous health management, onboard data processing
Maintenance	Limited, requiring complex missions	Minimal; failed satellites are replaced
End-of-life (EOL) management	Deorbiting or moving to graveyard orbits	Deorbiting; frequent technology refreshes
Reliability focus	High reliability and longevity	Scalability and adaptability
Communication	Continuous communication with ground stations	Reduced reliance on continuous ground communication
Private sector approach	Conservative, high-reliability methods	Scalable, flexible strategies
Aspect	Conventional spacecraft	Satellite constellations

© SAE International

1.2.
Historical Activities

Over three decades ago, NASA started a series of studies and workshops focused on spacecraft maintenance, a field that has evolved significantly since then. The late 1980s saw discussions on various aspects of servicing and support, culminating in the Satellite Services Workshop IV in June 1989 at NASA's Johnson Space Center.

The following are key points from past studies and workshops:

- The 1989 workshop showcased the strategic plan for spacecraft maintenance. It outlined NASA's vision for using the ISS, the Orbital Maneuvering Vehicle (OMV), and the Space Shuttle to achieve four primary goals: refueling or resupply, repairing, retrieving, and system upgrade.
- The 1988 report proposed using commercially available robotics systems for space

applications because of the limited investments in robotics technologies by NASA.

- NASA's strategy: The strategy proposed a progression from demonstrations in LEO to more complex operations in GEO and beyond. It highlighted the development of increasingly capable robotic systems and the establishment of a business model for space technology.
- A crucial aspect of the strategy was the need to enable common standards among satellite designs to support servicing, including modularity and commonality of connectors, ports, and grappling fixtures.
- The OMV program ended: Ironically, the OMV, a central element in NASA's servicing strategy, was discontinued in 1990 because of budget constraints. This resulted in a halt in high-level planning for satellite servicing.
- NASA Telerobotics Initiative: In the 1980s and 1990s, this initiative funded projects associated with on-orbit assembly and servicing, science payload tending, and planetary surface robotics.

The development of maintenance and support took a less direct path than predicted in past studies. However, it progressed through a combination of need, technological readiness, and the recognition of its benefits. The field witnessed notable progress, particularly in the areas of autonomous rendezvous, docking technologies, and robotic systems. The insights gained from past studies, combined with recent technological advancements, continue to shape the future of maintenance in space. This involves the potential for more sophisticated on-orbit repairs, refueling missions, and the extension of satellite lifespans, enhancing the utility and economic viability of space assets [1.5].

1.2.1.
Early Demonstrations

Spacecraft maintenance technology boasts a rich and varied history, stretching back to the earliest days of space exploration. From the first manned spacecraft to the latest innovative space missions, maintenance plays a crucial role in ensuring the success of each mission [1.6, 1.7]. In this section, we will examine the history of spacecraft maintenance technology, from its earliest beginnings to its current state and its potential future.

The first manned spacecraft missions were simple, short, and relatively low-tech affairs. Vostok 1 was started by the Soviet Union in 1961. It had one Earth orbit, with no plans for vehicle repair or maintenance. The same was true of the early US Mercury missions, which were similarly simple and short (Figure 1.4).

Figure 1.4 (a) The Vostok 3KA-3 spacecraft (Vostok 1) awaits the launch of Yuri Gagarin on April 12, 1961, which would make him the first human to travel into space. (b) The Soyuz spacecraft replaced the Voskhod and serves to ferry crew to or from the ISS.

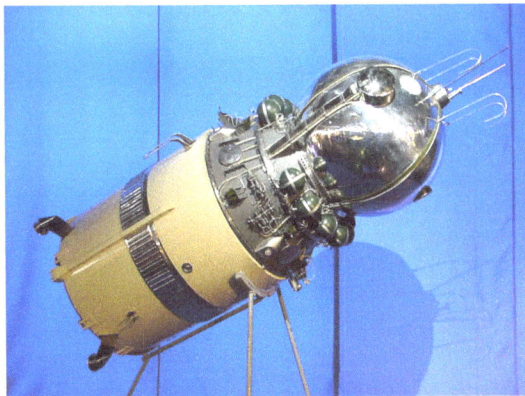 

(a)   (b)

It was not until the Gemini program that spacecraft maintenance assumed a more significant role. The Gemini spacecraft was created for longer missions, and as such, they needed to be repaired and maintained in space. These missions also presented the concept of spacewalking, which allowed astronauts to leave the spacecraft and perform maintenance tasks outside in the vacuum of space. These tasks involved hands-on servicing, which was the initial approach to satellite maintenance. Astronauts would perform satellite maintenance and repairs during missions in orbit. These missions were risky and expensive because of the high cost of launching and maintaining a manned spacecraft. Manual SMs were limited in their ability to address complex problems, as astronauts had to work within the constraints of their suits and tools.

In the 1970s, the Remote Manipulator System (RMS) was created by NASA to address the limitations of manual SMs. The RMS, a robotic arm, could operate from within the spacecraft or the ground to capture and manipulate objects. It was used extensively during the construction of the ISS. While this showcased a notable advancement in satellite maintenance technology, the RMS was still limited in its ability to address complex problems. A few years later, in 1975, the Soviet Union sent their first space station, Salyut 1, into space, which boasted a system for docking and undocking with visiting spacecraft. This enabled crew rotation and resupply, representing a significant milestone in the technology's development and capability for servicing and maintaining spacecraft in orbit.

The early demonstrations of satellite maintenance showed a significant evolution in space exploration, beginning with the launch of

Skylab, NASA's first space station, on May 14, 1973. Despite its tumultuous start and immediate technical issues postlaunch, Skylab's challenges spurred the development of on-orbit repair capabilities. The Skylab crew, sent into space just 10 days after the station, successfully conducted extravehicular activities (EVAs) to rectify the issues, particularly with the micrometeoroid shield and solar arrays. This mission not only recovered Skylab but also showed the feasibility and importance of on-orbit repair, setting a precedent for future space missions.

The 1980s signified a new chapter in the history of spacecraft maintenance. During a time of technological innovation and space exploration, extended human missions necessitated a skilled team of maintenance technicians to ensure the safety and functionality of shuttle vehicles for each mission. During this period, several experiments were conducted. However, the Solar Maximum Mission (SMM), a NASA satellite, holds the distinction of being the first orbiting spacecraft to undergo repairs while in space. The SMM was primarily designed to study solar flares.

Shortly after its launch, in November 1980, one fuse of the satellite experienced a failure, rendering it inoperable. Despite being inactive for four years, the SMM was eventually repaired by astronauts aboard the Space Shuttle. The SMM malfunction caused a multiyear delay, emphasizing the importance of timely maintenance and repair of space assets, especially in critical services such as geolocation and telecommunications. As a result, the focus on creating rapid solutions for prolonging the lifespan and repairing satellites kept growing. This incident highlighted the importance of robust and reliable maintenance protocols for space assets, which are critical for ensuring the continuity of essential services. Another Space Shuttle, Columbia, was sent off on April 12, 1981. This shuttle program executed a variety of tasks, including carrying satellites into orbit, conducting scientific experiments, and servicing the HST.

Not only was the Space Shuttle a new type of spacecraft, but it also introduced limited onboard maintenance technology, such as in-flight repairs, orbital debris mitigation, and inspection technology.

The Multimission Modular Spacecraft (MMS) design emphasizes modularity for easier repair and replacement in space, as well as reduced costs for ground integration and testing, advancing spacecraft technology. The first to make use of this design was the SMM, launched in 1980 to study solar phenomena. However, a failure in its attitude control system in 1981 ended the mission prematurely. In response, NASA started an ambitious recovery mission in 1984 with the Space Shuttle Challenger. The expedition successfully secured, repaired, and redeployed the SMM, fully restoring its operation and showing the potential of the Space Shuttle as a tool for on-orbit servicing.

After achieving success with SMM, the Space Shuttle was again pivotal in another on-orbit servicing challenge involving the Palapa B2 and Westar 6 satellites. Both satellites had issues after deployment from the shuttle Challenger. In a dramatic mission in November 1984, the Discovery STS-51A mission recovered these satellites, showcasing the shuttle's capability in on-orbit servicing. This mission was not only a technical achievement but additionally a commercial triumph, as it showcased the economic benefits of satellite servicing. The satellite insurance industry acknowledged this

capability, recognizing the potential for on-orbit servicing to mitigate losses from satellite failures.

There were further difficulties during this period. On January 28, 1986, the Space Shuttle Challenger suffered a catastrophic failure during launch, killing all seven crew members on board. The disaster resulted from a failure in one of the shuttle's powerful rocket boosters, which led to the booster's separation from the shuttle's main fuel tank. The Challenger disaster was a pivotal moment as it led to a redesign of the propulsion system and a renewed focus on safety in space travel. However, it further indicated the risks and dangers associated with space exploration, reminding us that even with the best technology and maintenance, accidents can still (and will) happen.

The early missions—Skylab, SMM, Palapa B2, and Westar 6—collectively showed that on-orbit repair and refurbishment could significantly extend the life and functionality of space assets. They highlighted the strategic and commercial value of satellite servicing, paving the way for its integration into future space mission designs. This era in space exploration set in motion on-orbit servicing as a critical component, transforming how we approach the design, operation, and sustainability of space missions.

In the 1990s, NASA's HST received maintenance and upgrades on four separate missions between 1993 and 2009. HST was creatively engineered for on-orbit servicing, revolutionizing the field with its approach to crew training, tool design, procedure testing, and verification. The mantra of "test, test, and retest" and "train, train, and retrain" played a crucial role in ensuring the success of complex SMs. HST's SMs varied from straightforward box replacements to intricate board-level repairs, showcasing the evolution of the Orbital Replacement Unit (ORU) and Orbital Replacement Instrument (ORI) design. During missions to HST, astronauts executed delicate instrument replacements and repairs.

Engineered for servicing: HST was a ground-breaking project, but particularly in its construction for on-orbit servicing. This foresight enabled the replacement of instruments and the correction of unforeseen issues, like the spherical aberration in its primary mirror. Sent into space on April 25, 1990, by the Space Shuttle Discovery, HST faced challenges like a flawed primary mirror and shaking from its solar arrays. The blurring of observatory images caused reputation damage to the agency. The Corrective Optics Space Telescope Axial Replacement (COSTAR) and the Wide Field and Planetary Camera 2 (WFPC2) were designed to correct the telescope's vision. The European Space Agency (ESA) also took part by rebuilding improved solar arrays to minimize thermal sensitivity. Over its lifetime, HST experienced five SMs. Each mission resulted in substantial upgrades and repairs.

- SM1 (1993): Rectified the mirror flaw with COSTAR and WFPC2, replaced solar arrays, and performed other critical repairs. Over five consecutive days of EVAs, the crew successfully executed all planned activities to fully restore all capabilities and even enhance them in some respects. This mission established a precedent for future SMs, demonstrating the utility and versatility of satellite servicing.

- SM2 (1997): Implemented sophisticated instruments like the Space Telescope Imaging Spectrograph and the Near Infrared Camera and Multi-Object Spectrometer, enhancing HST's scientific capabilities.

- SM3A (1999) and SM3B (2002): Resolved failing gyroscopes, installed new instruments like the Advanced Camera for Surveys, and made significant HW upgrades.

- SM4 (2009): The final maintenance mission, which installed new instruments (Cosmic Origins Spectrograph and Wide Field Camera 3), replaced gyroscopes and batteries, and added a soft capture mechanism for future deorbiting.

The SMs not only extended HST's operational life, but also improved the telescope's scientific capabilities by maintaining its operational state. The ability to enhance HST with the latest technology turned it into one of the most significant scientific instruments in history (Figure 1.5).

Figure 1.5 Three crewmembers actively engaged in Intelsat 603 during the STS-49 mission.

NASA.

Finally, a notable development during this period was the OMV. It was among the first robotic servicing systems created to move payloads and re-boost facilities near a space station. Unfortunately, the OMV was not technologically mature enough, and its high cost was a significant drawback. Despite being a key component of the 1980s designs for the spacecraft, NASA terminated the program in 1990 because of budget pressures and the lack of immediate requirements. Many credit the cancelation of the mission to NASA's "faster, better, cheaper" philosophy. This philosophy was based on the belief that it would cost less to replace the target satellite with a robot rather than sending astronauts. However, this perspective was short-lived as the manually handled system components were not as robust as expected.

As technologies that facilitate robots' advancements progressed significantly with several demonstrations, it highlighted the significance of adapting human spaceflight HW to service goals. For instance, the STS-49 mission to Intelsat IV satellite 603 in May 1992, the STS-61 "rescue mission" of HST in 1993, and the STS-82 maintenance mission to HST in 1997 demonstrated the necessity of assisted maintenance for satellites. The progress in robot technology, with enhanced abilities and capabilities, paved the way for more efficient and cost-effective spacecraft servicing. In hindsight, the OMV program acts as a clear reminder of how challenging it is to develop and execute a groundbreaking idea. Yet, it also serves as evidence of the ingenuity and resilience of scientists and engineers who persistently work toward the advancement of space maintenance technology.

1.3.
The Ever-Rising Importance of Maintenance

Finally, in 2007, the US Defense Advanced Research Projects Agency (DARPA) started the Orbital Express mission. This mission intended to showcase a reliable and cost-effective approach to maintaining satellites in orbit with a new feature: autonomously. Two spacecraft prototypes played a role in the mission: NEXTSat, created by Ball Aerospace, and ASTRO, designed by Boeing. ASTRO was initially planned as a maintenance and repair spacecraft, unlike the NEXTSat, which was specifically designed as a modular and serviceable satellite. Initially, the task aimed to conduct tests. In addition, the mission aimed to showcase the use of advanced algorithms and sensors for automated rendezvous and docking maneuvers. A noteworthy milestone was reached in the development of automated and cost-effective in-orbit servicing with the overall mission.

Although ground control intervened twice, DARPA views this mission as a pivotal moment for the industry. It demonstrated the possibility of autonomous satellite servicing and the potential for substantial cost reductions. The success paved the way for further research and development in automated satellite maintenance in the space industry.

Even with the success of the Orbital Express mission, the commercial satellite industry still encountered significant challenges in adopting this approach because of several technological, regulatory, and financial barriers. Before automated in-orbit maintenance can be implemented, the industry wanted further:

- Technological advancements: One major technological barrier was the nonexistence of standardized interfaces and connectors on satellites, which made it difficult for servicing spacecraft to dock with and manipulate satellites. The complicated and varied design of satellite components posed a challenge in developing robotic systems that could perform the maintenance tasks. The development of autonomous systems capable of functioning in the harsh environment of space, including extreme temperatures, radiation, and vacuum, presented a significant technical challenge. Also, the development of standardized interfaces and connectors for satellites will be crucial. Additionally, the creation of robotic systems capable of executing complex maintenance tasks in space will play a vital role in facilitating widespread adoption.

- Regulations: The legal and regulatory framework that oversees space activities was primarily designed to regulate launches and reentries, rather than on-orbit activities such as satellite servicing. This shows that there was a lack of clear rules and standards governing in-orbit servicing, thereby creating difficulties for companies to plan and execute satellite SMs. Moreover, unclear liability and responsibility during maintenance operations deterred investors from funding missions.

- Financial support: The high costs of developing complex maintenance robots for space, along with the expenses of launching and operating servicing spacecraft, made it hard for companies to justify the investment in financial support. The absence of a clear business case for satellite servicing posed challenges for organizations to secure financing for these missions.

Despite these barriers, the significance of research and development in enhancing the technology and capabilities of satellite maintenance and repair cannot be emphasized enough. As the commercial space industry keeps growing, the need for reliable and efficient methods of satellite maintenance becomes increasingly important. In parallel with the HST missions, the ISS emerged as an important platform for developing satellite maintenance technologies and techniques. The assembly necessitated many spacewalks and showed the value of standard interfaces and modular design for on-orbit connection. This was a massive demonstration of on-orbit construction and maintenance that involved many EVAs and robotic operations, setting the stage for future large-scale space construction and maintenance projects.

Robots have had a substantial impact on ISS construction and maintenance. The Space Station Remote Manipulator System (SSRMS) and the Canadian Special Purpose Dexterous Manipulator (SPDM), or Dextre, have facilitated various operations, previously requiring spacewalks to be performed robotically. These systems have since been employed for tasks ranging from capturing visiting vehicles to replacing components in the ISS itself. Dextre's capabilities are being expanded through demonstrations like the Robotic Refueling Mission (RRM) and the Dextre Pointing Package (DPP), which aim to advance autonomous rendezvous, capture techniques, and fine-pointing capabilities for Earth and space instruments. Robonaut 2 (R2) is another human-equivalent robot that improves the ISS's robotic capabilities. Initially stationed inside the ISS, R2 eventually assists in external tasks, furthering the development of astronaut aids for space maintenance.

Experts in the field agree that the existence of the ISS, continuously occupied by astronauts since November 2000, has greatly advanced spacecraft maintenance technology. Regular maintenance and repairs are necessary to keep it functioning. The station has a variety of robotic and manned systems for performing maintenance tasks, including robotic arms, spacewalks, and specialized tools. Besides the space station, many other spacecraft and satellites necessitate maintenance and repairs. Astronauts manage some missions, while robotic systems maintain others. The Mars rovers, for example, were outfitted with tools and instruments for performing basic maintenance tasks, such as cleaning their solar panels and clearing away dust. Progress in advancing the technology strategy since the NASA Exploration Team report in 2002 has been limited. However, significant efforts by other agencies, including the US, Europe, Japan, and elsewhere, have highlighted the potential for this capability [1.8].

As space technology becomes increasingly important in our daily lives, the need for efficient solutions to extend the lifespan of satellites in orbit has become urgent. The space industry's rapid growth and evolution have led to an increasing need for autonomous and efficient maintenance and repair methods for space assets.

1.4.
The Scope of This Book

In the past 20 years, we have seen significant advancements with reusable spacecraft, robotic arms, and autonomous technology, to name a few. Spacecraft health management design has changed, as astronauts used to service shuttles in space (or not at all). During this period, many

ideas were considered for maintenance and repair, often with flaws in their design. Yet, progress remained evident in highlighting the role of technology. Effective maintenance has become an integral part of spacecraft design, guaranteeing the long-term success of space missions. We understand it encompasses a wide range of activities. Our goal is to extend the operational lifespan, reduce mission failure, and ensure crew and payload safety. In this book, we will explore the technology supporting this agenda, challenges engineers face when working with spacecraft, and future perspectives on health management technology.

The book chapters are separated into two parts. The first part, Chapters 1–5, focuses on fundamentals and systems. The second part, Chapters 6–14, focuses on advanced concepts and applications.

Chapter 1 examines the evolution of the space industry, highlighting the transition from government-led to private-sector involvement. It explores the impact of this shift on innovation, cost reduction, and the broader accessibility of space. The aim is to lay the groundwork for comprehending the current landscape of space technology and the importance of SHM in this evolving context.

Chapter 2 provides an overview of the different subsystems within a spacecraft. It establishes the foundational knowledge necessary to appreciate the complexities of spacecraft operation and the critical role of health management.

Chapter 3 underscores the significance of **proactive health management** in preventing mission failures and ensuring safety, drawing lessons from previous missions. This is examined by building upon the previous chapters and emphasizing the need for robust health management systems amid the complexities of spacecraft systems.

Chapter 4 covers techniques for managing and diagnosing faults within spacecraft systems, which include FTA and model-based reasoning. It provides an insight into the methods used to identify and mitigate potential failures, a key component of health management.

Chapter 5 examines the specifics of spacecraft health management and emphasizes preemptively addressing system health issues through the integration of sensor networks and data analysis. It elaborates on the concept of health management by detailing the technologies and methodologies employed to maintain spacecraft integrity.

Chapter 6 investigates how telemetry data can help maintain the reliability of spacecraft operations, outlining the process of detection and diagnosis in detail. It ties into the theme of technology-driven decision-making in health management, showcasing the practical application of telemetry data.

Chapter 7 examines the use of data-driven approaches for analyzing spacecraft data to predict and prevent failures. It underscores the advanced analytical techniques that represent the cutting edge of health management strategies.

Chapter 8 explores how probabilistic design and various risk assessment methods are used to enhance the reliability and operability of a spacecraft within the constraints of tighter budgets and shorter development schedules.

Chapter 9 centers on onboard processing and its challenges in spacecraft design, addressing the challenges and opportunities presented by AI. This is achieved through discussions on the

technical and design considerations necessary to implement the health management strategies outlined in previous chapters.

Chapter 10 investigates various concepts, such as cybersecurity, SW security, and human–machine interfaces (HMIs), emphasizing their role in comprehensive health management. It expands the scope of health management to include cybersecurity and human factors, illustrating the multidisciplinary nature of maintaining spacecraft health.

Chapter 11 addresses the future of in-space servicing and maintenance, which considers the technological/economic implications and expects the practical application and evolution of health management practices in space servicing.

Chapter 12 discusses the potential of DTs in spacecraft design and health management. It presents a futuristic perspective on simulation-driven development and maintenance. The technology has potential to revolutionize health management by providing a virtual mirror of physical spacecraft for real-time monitoring and simulation.

Chapter 13 explores the role of autonomy in spacecraft health management, detailing the state-of-the-art solutions and processes that enable independent system monitoring and decision-making. It wraps up the discussion by envisioning a future where spacecraft health management systems operate autonomously, leveraging the concepts and technologies explored in previous chapters.

Chapter 14 presents the author's insights into the future of AI-based systems, robotic autonomy, and the technological roadmap for spacecraft health management. It acts as a culmination that ties together the themes of the book, offering a reflective look at the future directions of spacecraft health management technology.

The chapters collectively develop a comprehensive narrative on the evolution, current state, and future of spacecraft SHM. Starting with the backdrop of the space industry's evolution, the book explores the technical complexities of spacecraft systems, the criticality of health management, and the emerging technologies that will shape its future. Each chapter develops upon the previous ones, layering detailed technical insights with broader thematic connections on integrating AI, autonomy, and DTs. This progression shows a logical flow from historical context and system fundamentals to recent innovations, illustrating a holistic view of the field's past, present, and potential future.

References

1.1.   Iacomino, C., "Evolution of Innovation Mechanisms to Support the Post-2030 Agenda Goals: Case Study on the European Space Exploration Programme," in Froehlich, A. (ed.), *Post 2030-Agenda and the Role of Space*, Studies in Space Policy (Cham: Springer, 2018), doi:https://doi.org/10.1007/978-3-319-78954-5_9.

1.2.   Khan, S., "Redefining Space Commerce: The Move toward Servitization (No. EPR2024002)," 2024.

1.3.   Horsham, G.A.P., Schmidt, G., and Gilland, J., "Building a Robotic, LEO-to-GEO Satellite Servicing Infrastructure as an Economic Foundation for 21st-Century Space Exploration," *International Journal of Space Technology Management and Innovation* 1 (2011): 1, doi:https://doi.org/10.4018/IJSTMI.2011010101.

1.4.   Johnshoy, M., "The Final Frontier and a Guano Islands Act for the Twenty-First Century: Reaching for the Stars without Reaching for the Stars," March 22, 2012.

1.5.   Nayar, H., Ali, K., Aubrey, A., Estlin, T. et al. "OTC 22989-PP Space Robotics Technologies for Deep Well Operations," Paper Presented at in *Offshore Technology Conference*, Houston, 2012.

1.6.   Davarian, F., "Prolog to the Section on Space Exploration and Science," *Proceedings of the IEEE* 100 (2012): 1782-1784, doi:https://doi.org/10.1109/JPROC.2012.2187134.

1.7.   Hasan, R. and Hasan, R., "Towards a Threat Model and Security Analysis of Spacecraft Computing Systems," in *2022 IEEE International Conference on Wireless for Space and Extreme Environments (WiSEE)*, Winnipeg, MB, Canada, 2022, doi:https://doi.org/10.1109/WiSEE49342.2022.9926912.

1.8.   McKenzie, P.M. and Gilbert, C., "International Industrial Cooperation: The Key to an Affordable and Sustainable Space Exploration Program," in *1st Space Exploration Conference: Continuing the Voyage of Discovery*, Orlando, 2005, doi:https://doi.org/10.2514/6.2005-2530.

The study of spacecraft systems is an essential interdisciplinary pursuit that bridges the fields of engineering, science, and external phenomena. As such, some knowledge of the basic principles and core technologies is important. Typically, available texts on the subject area either oversimplify and lack depth or cater exclusively to professionals with highly technical material. I have added this chapter to summarize some essential concepts in modern spacecraft. The purpose is to give readers the necessary vocabulary and understanding for easier engagement with the book's content. I intend to discuss the scientific and engineering advancements of spacecraft systems, encompassing the various interactions within the spacecraft as well as those with its external environment. This should serve as a valuable introduction for readers.

A spacecraft operates in space and can be designed for either suborbital or orbital spaceflight. In suborbital spaceflight, the spacecraft travels into outer space but returns to the Earth's surface without completing a full orbit. On the other hand, an orbital spaceflight involves the spacecraft entering a continuous orbit around a celestial body. For human spaceflights, crew or passengers are carried by spacecraft, which serve a variety of purposes such as communication, EO, meteorology, navigation, planetary exploration, and space tourism (   ). These themes are also popular in science fiction, sparking the imagination of many.

Figure 2.1 Apollo 15 Command and Service module. It consisted of the conical command module, a cabin that housed the crew and carried equipment needed for atmospheric reentry and splashdown.

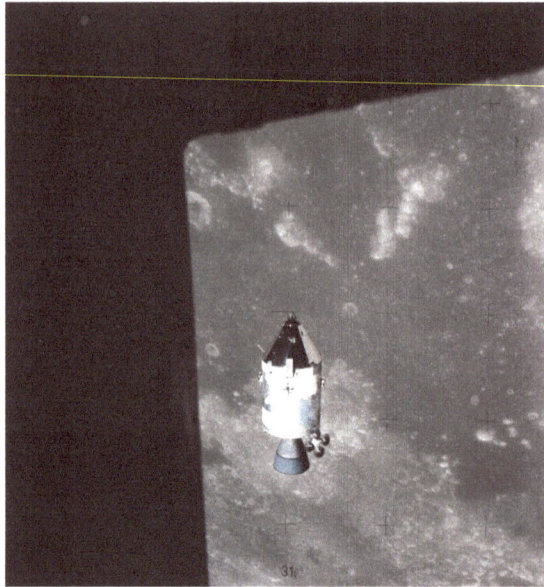

Spacecraft consists of a diverse array of subsystems, carefully tailored to meet specific mission requirements and payload needs [2.1]. These subsystems cover critical functionalities that are essential for smooth operation. One such crucial subsystem is **attitude determination and control (ADC)**, which accurately establishes and maintains the orientation of the vehicle in space and ensures precise maneuverability and control. Another vital subsystem is **guidance, navigation, and control (GNC)**, which has a pivotal role in guiding the spacecraft's path through the vastness of space. With the installation of advanced algorithms and sensors, this subsystem enables accurate navigation and control throughout the mission [2.2, 2.3]. The **communications subsystem (COMS)** creates and maintains communication links between the spacecraft and mission control on Earth. It facilitates the transmission of essential data, commands, and scientific observations, allowing real-time monitoring and control of the mission. Managing commands and organizing data is the **command and data handling (CDH) subsystem** [2.4]. Serving as the information hub of the spacecraft, this subsystem collects, stores, and processes data while executing commands from ground control. To fulfill the power requirements of the spacecraft, the **power subsystem** ensures the generation, distribution, and regulation of electrical power. It combines different power sources, such as solar panels or batteries, to provide the energy for all onboard systems. Ensuring optimal temperature and protecting sensitive components is the role of the thermal control subsystem. This particular subsystem actively oversees the spacecraft's thermal conditions, shielding it from extreme temperatures and ensuring the reliable performance of critical systems. **Propulsion**, a key aspect of spacecraft operation, is enabled by the propulsion subsystem. It relies on engines and propellants to propel the spacecraft, adjust its trajectory, and enable orbital maneuvers [2.5]. Offering mechanical support and structural integrity is the **structure** [2.6]. It is engineered to endure the demanding conditions of launch, the harsh environment of space, and the dynamic forces encountered during mission operations. Together, these interconnected subsystems create a comprehensive spacecraft system, working in harmony to enable successful space missions.

The systems include a wide range of spacecraft, each serving distinct purposes and fulfilling specific mission objectives. We can categorize these into different groups based on their design and intended applications. Some of the commonly acknowledged types include:

- Satellites: These are an important type of spacecraft that fulfill various tasks such as communication, navigation, weather forecasting, and scientific research. Typically, we position them in orbit around Earth, where they maintain a stable position relative to our planet. Satellites can capture pictures, monitor the weather, and study nearby stars and other celestial bodies. Since 1990, the HST, one of the most renowned types of satellites, has been in orbit and continues to make significant contributions to our understanding of the cosmos. Chapter 1 highlights the HST as a demonstration of the potential of satellites and its critical role in advancing our knowledge.

- Probes: These are unique spacecraft that are specially designed to explore nearby planets and other celestial bodies in space. These probes travel from Earth through space to their destination, where they can gather data and send them back to us. Some of the most famous probes comprise the Voyager probes, which have been exploring the outer reaches of our solar system since their launch in 1977. These probes have given invaluable data about gas giants and their moons. Another notable probe is the Mars Curiosity rover, which touched down on Mars in 2012. Since then, it has continued to collect data about the planet's climate, geology, and atmosphere.

- Space shuttles are reusable spacecraft used to transport people and cargo to and from Earth. They provide a budget-friendly and eco-friendly option for launching and retrieving payloads, which makes them perfect for constructing and supplying resources to space stations like the ISS. However, building and keeping up with space shuttles can be costly and require some clever engineering to implement. In 2003, losing Space Shuttle Columbia and her crew during reentry showed the tragic dangers of human spaceflight. Despite these risks, shuttles continue to be an important tool for space discovery and work.

- Space stations: These types of spacecraft allow humans to live in space. Since it was launched in 2000, the ISS has been a center for innovative research and global collaboration. It brings together nations in the spirit of exploration. However, maintaining such a structure in space poses unique challenges that demand constant maintenance and repair to ensure safety and research mission success.

2.2.
ADC

ADC is responsible for precisely establishing and maintaining the spacecraft's orientation in space. A spacecraft in space must align toward a direction as assigned by the mission requirements. Many assets are Earth-oriented while others are inertial space object-oriented, such as the Sun or an asteroid/star of interest. We understand this orientation as attitude. To achieve control and stabilization, attitude sensors are employed to determine the current attitude, and actuators are employed to generate the required torque to maintain or modify the current attitude and carry out its intended tasks. The subsystem uses a variety of sensors to gather data on the spacecraft's position, rotation rates, and orientation relative to celestial objects and the Earth's magnetic field. These can comprise gyroscopes, star trackers, Sun sensors, magnetometers, and horizon sensors. Attitude control requires the application of control torques to adjust the spacecraft's attitude and maintain the desired orientation. Engineers employ actuators such as reaction wheels, thrusters, magnetic torques, and control moment gyroscopes to generate these torques. The control system employs feedback from the attitude determination sensors to calculate the control commands, which are then sent to the actuators. By appropriately changing the torques, the spacecraft can counteract external disturbances, maintain stability, and perform precise maneuvers.

ADC systems have been employed in various spacecraft missions. One significant early demonstration was in the Mercury spacecraft during the Mercury program, which successfully brought astronauts to space in the 1960s and 1970s. The Apollo spacecraft used gyroscopes, accelerometers, and thrusters to determine and control its attitude. These systems were essential for preserving the spacecraft's stability during vital mission phases such as lunar landings and reentry into Earth's atmosphere.

Another significant milestone in ADC was the deployment of the Skylab space station by NASA in 1973. Skylab included advanced ADC systems to ensure precise pointing of its solar panels and instruments toward the Sun. The ADC subsystem used gyroscopes, Sun sensors, and reaction wheels to precisely manage Skylab's orientation, facilitating efficient power generation and scientific observations. Other space observatories and satellites, such as the Kepler Space Telescope and the ISS, also use advanced ADC systems. These systems facilitate precise pointing, stabilization, and maneuverability to carry out their respective missions. The HST made significant contributions to astronomy and possessed an advanced ADC system. This spacecraft is one of many that improved ADC technologies over time. Each mission had built upon the lessons learned from previous endeavors, continually advancing the technology.

2.3.
GNC

GNC systems facilitate precise and controlled movement in space. These systems have a vital function in ensuring the spacecraft's ability to navigate accurately, maintain its desired trajectory, and perform necessary maneuvers [2.7].

- Guidance involves establishing the spacecraft's intended path or trajectory. It requires the use of algorithms and computations to calculate the control inputs to achieve specific objectives, such as reaching a target destination or maintaining a stable orbit. The guidance system considers various factors, including mission requirements, spacecraft capabilities, and external influences, such as gravitational forces and atmospheric drag. By offering guidance commands, the system directs the spacecraft along its intended course.

- Navigation requires the determination of the spacecraft's position, velocity, and orientation in space. Accurate navigation is crucial for upholding the spacecraft's desired trajectory, avoiding collisions with other objects, and enabling precise targeting of destinations. Navigation systems rely on various sensors, including star trackers, gyroscopes, accelerometers, and Global Navigation Satellite System (GNSS), to collect data on spacecraft motion and position in relation to reference points. Then, experts examine and evaluate this information to provide real-time updates on the spacecraft's location and velocity.

- Control is the capacity to handle the spacecraft's attitude, position, and velocity in order to accomplish the desired mission objectives. Control systems make use of various actuators, such as reaction wheels, thrusters, and control moment gyroscopes, to apply forces and torques on the spacecraft. This allows for accurate adjustments in the spacecraft's orientation and motion. Control algorithms and feedback loops, guided by engineers, continuously monitor and adjust the spacecraft's state using navigation data and guidance commands. The control system guarantees stability, responsiveness, and maneuverability, enabling the spacecraft to perform complex tasks such as orbit insertion, station keeping, and attitude stabilization.

Combining GNC systems enables spacecraft to perform a wide range of missions with precision and reliability. Whether it is a satellite in geostationary orbit, a planetary probe investigating distant worlds, or a crewed spacecraft en route to the ISS, GNC systems are fundamental to their success. Advancements in GNC technologies have revolutionized space exploration, enabling more accurate and autonomous spacecraft operations. Modern GNC systems integrate sophisticated algorithms, advanced sensors, and intelligent control strategies. These systems adapt to evolving mission requirements, withstand environmental disturbances, and optimize performance and efficiency. The successful implementation of GNC systems relies on extensive testing, simulation, and validation to confirm their performance and ensure mission success. Engineers and scientists continually strive to refine and improve GNC algorithms, sensor technologies, and control strategies to enhance the capabilities of spacecraft and enable ambitious missions. The Apollo Lunar Module (LM) can be one of the greatest and most popular examples of GNC. While it may not be the first demonstration, it is notable for its significance and technical achievements (Figure 2.2).

Figure 2.2 The Apollo LM. This module was transported to lunar orbit attached to the Apollo command and service module. It was the first manned spacecraft to operate exclusively in the airless vacuum of space and remains the only vehicle with humans on board to land anywhere beyond Earth.

NASA

During the Apollo Moon missions, the LM had a crucial role in safely landing astronauts on the lunar surface and returning them to the Command Module in lunar orbit. The GNC system controlled the spacecraft during complex maneuvers using onboard computers, sensors, and navigational aids. It determined the spacecraft's position and guided it to the intended landing site. The navigation system of the LM incorporated inertial measurement units, which used accelerometers and gyroscopes to measure the spacecraft's motion and attitude. It also integrated radar systems that offered accurate altitude and range measurements during the descent and landing phases. The control system of the LM used reaction control thrusters and descent engines to maneuver and stabilize the spacecraft. The guidance computer and pilot inputs controlled these propulsion systems to guarantee precise control over the spacecraft's attitude and motion.

Today, GNC systems remain essential in various space missions, including satellite operations, planetary exploration, and crewed spaceflight. These setups have grown more sophisticated, incorporating advancements in computer processing, sensor technologies, and autonomous control algorithms.

2.4.
COMS

The COMS facilitates the exchange of information between/among spacecraft and ground. It plays a crucial role in facilitating data transmission, command reception, and telemetry acquisition, establishing a vital link for space missions. It sets up and maintains communication links, ensuring reliable and efficient transmission of data, including images, telemetry, and command instructions. The following are key elements of the COMS:

- Transmitters and receivers: The COMS incorporates transmitters to send signals from the spacecraft to ground stations or other receivers in space. Receivers capture and interpret signals received from external sources.

- Antennas: They function as the interface between the spacecraft and the external environment for signal transmission and reception. They send and receive electromagnetic waves efficiently, allowing for long-range and reliable communication.

- Modulation and demodulation: Modulation techniques are used to encode information onto the carrier signals for transmission. Demodulation is employed to extract the original information from the received signals. Various modulation schemes are implemented depending on factors such as data rate, signal strength, and interference resilience.

- Data encoding and decoding: The COMS uses encoding techniques to convert raw data into a format suitable for transmission. This guarantees data integrity and minimizes errors during transmission. The recipient end performs decoding to retrieve the original data from the received signals.

- Communication protocols: These establish the rules and standards for data transmission and control within the COMS. They control aspects such as data formatting, error detection and correction, synchronization, and flow control.

- Telemetry and telecommand: The COMS enables the transmission of telemetry data from the spacecraft to ground stations, providing essential information about the spacecraft's health, status, and scientific measurements. It also allows telecommand signals from ground stations to control and configure spacecraft operations.

- Signal processing: The COMS utilizes signal processing techniques to enhance the quality and reliability of received signals. This involves techniques such as error correction coding, signal filtering, and noise reduction.

A notable milestone in COMS was the successful transmission of the first images of Earth from space during the early Mercury and Gemini missions of the 1960s. These missions utilized radio waves to establish communication links between the spacecraft and ground stations, allowing for the transmission of vital data, such as images, back to Earth. These early demonstrations of COMS established the groundwork for future developments in space communication. Other accomplishments resulted from the Apollo missions in the late 1960s and early 1970s, which involved the first human landing on the Moon. These missions heavily relied on creating and maintaining communication links over vast distances. Since achieving those early milestones, there have been many advancements in space communications. These advancements include the development of deep space communication networks, like NASA's Deep Space Network (DSN), which allow for continuous communication with spacecraft exploring the outer regions of our solar system. Advanced modulation techniques, error correction coding, and signal processing have greatly improved data transmission in space. This benefits various applications such as telecommunications, broadcasting, weather monitoring, and navigation systems.

2.5.
CDH

The CDH subsystem has a crucial role in the execution and management of mission operations. It includes the HW and SW systems that accept commands from ground control, process and execute those commands, collect and store data from onboard sensors and instruments, and transmit data back to Earth for analysis. It acts as the "brain" of the spacecraft, guaranteeing proper functionality, data handling, and communication throughout the mission. The primary function of the CDH is to accept and execute commands from mission operators on the ground. These commands offer instructions for spacecraft operations, such as attitude control, instrument activation, data acquisition, and payload deployment. The system decodes and interprets these commands, guaranteeing accurate and safe execution. It organizes the activities of other subsystems, such as the ADC and the power system, to carry out the desired actions.

Data handling is another crucial aspect of the CDH subsystem. Since Spacecraft produces vast amounts of data from various sensors, instruments, and systems onboard, the CDH manages these data, including its collection, storage, processing, and transmission. The CDH guarantees that it organizes, formats, and stores the data reliably, making them ready for analysis and retrieval. Additionally, the system manages the optimization of storage capacity and ensures data security during transmission, leveraging a combination of HW and SW components to carry out its functions. A central processing unit (CPU) or flight computer is usually included in the HW components. It is accompanied by memory modules for data storage, input/output interfaces for communication with other subsystems, and interfaces for external connections with ground-based systems. The embedded SW includes the SW components that control the operations of the subsystem, manage data processing and storage, and handle the execution of commands and control algorithms.

2.6.
Power Source

This subsystem supplies electrical power to support mission operations and ensure the functionality of other spacecraft systems, including payloads and instruments, enabling them to perform tasks. Power subsystems are also responsible for generating, regulating, distributing, and storing electrical power throughout the mission. These would typically require one or more power generation technologies that also have to be regulated to different voltages required by specific systems and payloads. To guarantee continuous power availability during the mission, we primarily generate power from solar panels to convert sunlight into electricity. Additionally, we include energy storage devices such as batteries or capacitors that can store excess electrical energy generated during periods of peak solar illumination and discharge it when needed. Power sources can involve the following:

- **Batteries:** They function as a useful power source for short-term applications, such as during launch or docking maneuvers, where a constant power source is needed. However, because of their limited lifespan, batteries are not suitable for long-term missions. Instead, people often use them in conjunction with other power sources, such as solar panels, to offer a backup power supply. In this configuration, the batteries are used to store excess power produced by the solar panels during periods of high solar irradiance. For example, the batteries employed in the ISS utilize a mix of solar panels and batteries to generate and store power. It then utilizes this stored energy during times when it is in Earth's shadow or when the solar panels are not generating enough power.

- **Photovoltaic cells:** These are solid-state devices constructed with semiconducting materials that generate electricity when exposed to sunlight. They consist of two layers: a p-type semiconductor, which has an excess of positively charged holes, and an n-type semiconductor, which has an excess of negatively charged electrons. When sunlight strikes the cell, it excites the electrons in the n-type layer, causing them to move to the p-type layer, creating a voltage difference between the two layers. This voltage difference generates a direct current, which can power electrical systems. Since it is renewable and an abundant source of energy, it completely avoids producing any emissions or waste.

It also reduces the need for bulky fuel cells, which are regularly employed with batteries. By removing this requirement, we decrease the overall weight of the spacecraft, making it more economical to launch and operate. Most applications of photovoltaic cells are common in LEO missions, such as satellites used for transmitting information and monitoring weather. Additionally, researchers have employed them in the ISS, which depends on solar power to supply electricity to its systems and crew. Photovoltaic modules also play a major role in various robotic missions, such as the Mars rovers.

- **Radioisotope Thermoelectric Generators (RTGs):** These can produce electricity through the heat produced by the natural decay of a radioactive isotope, typically plutonium-238. An array of thermocouples crafted from silicon–germanium junctions then convert the heat into electricity. Unlike solar panels, which depend on sunlight, an RTG does not necessitate external factors to generate power, making it a dependable option for deep space missions. RTGs have several advantages over other power sources. They provide consistent power, regardless of external factors like sunlight or atmospheric conditions. Additionally, their longer lifespan makes them ideal for missions that require long-term power sources. However, a major concern is the potential danger presented by the radioactive material used to generate heat. There are many instances throughout the history of RTGs being employed for spacecraft, including the Voyager and Pioneer missions, which explored the outer reaches of our solar system. The Mars Curiosity rover also relies on an RTG for its power source.

- **Fuel cells:** These produce electricity using chemical reactions. The substances involved are typically hydrogen and oxygen, along with water as a by-product. Since fuel cells are an efficient and reliable power source, ensuring a constant supply of electricity with minimal maintenance required, they are used on manned spacecraft. Fuel cells have many benefits over other power supply systems. They are lightweight, have a high energy density, and generate no harmful emissions. Some instances involve the use of proton exchange membrane (PEM) fuel cells, which are commonly used in spacecraft and aircraft because of their high efficiency and low operating temperatures. Other types include the solid oxide fuel cell (SOFC), which operates at high temperatures, making it ideal for remote power generation, and alkaline fuel cells (AFCs), which are reliable and cost-effective but have low power density, limiting their suitability for space missions. However, since fuel cells necessitate a constant supply of fuel, it can limit their use on long-duration missions. They can also be relatively complex and require careful management to avoid the accumulation of dangerous gases. During the STS-83 mission, the Space Shuttle Columbia had to come back to Earth early because of a problem with its fuel cell.

The power subsystem monitors power consumption, allocates power based on mission requirements, and manages usage for longevity and stability. These mechanisms allow the spacecraft to function within the available power constraints and adjust to varying power demands during different mission phases (Figure 2.3).

Figure 2.3 (a) The TEMPEST-D spacecraft with its solar panels deployed. This spacecraft is part of a mission to experiment with new technologies for predicting weather and understanding storms. (b) NASA outfitted the Voyager probes with three RTGs that convert heat generated by the decay of plutonium-238 into electricity, providing power for the spacecraft.

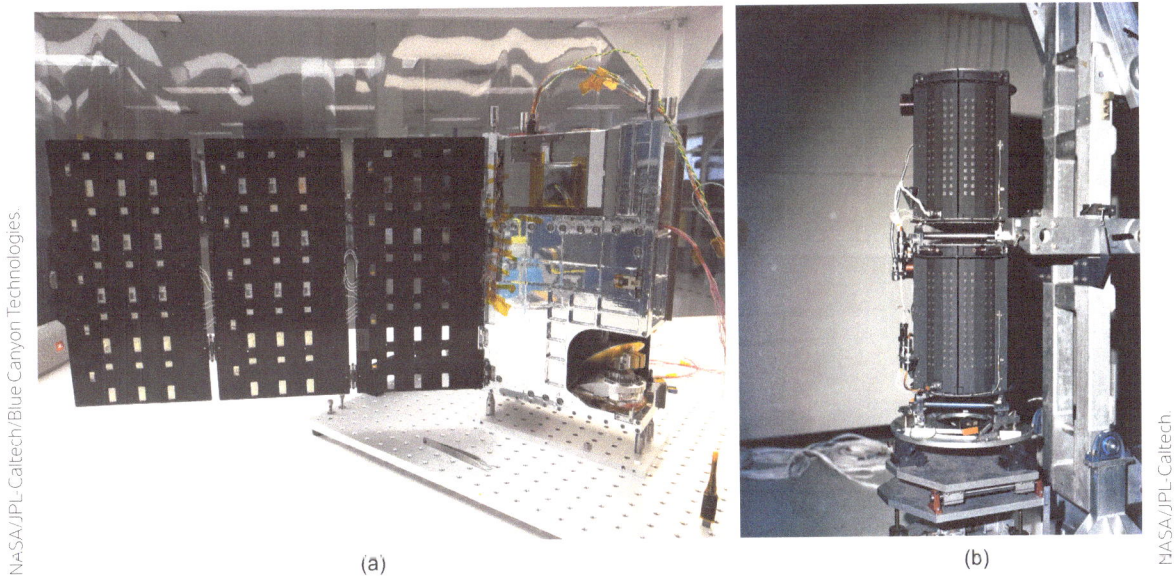

NASA/JPL-Caltech/Blue Canyon Technologies.

(a)

(b)

NASA/JPL-Caltech

2.7.
Thermal Control

The thermal control subsystem manages and maintains the temperature of various systems and components within the spacecraft. It plays a critical role in ensuring the proper functioning and longevity of onboard equipment, as well as the comfort and safety of crew members in human spaceflight missions. Space is an extreme environment with significant temperature variations, from extreme cold to intense heat based on location and solar radiation impact. These temperature extremes can have detrimental effects on sensitive electronic components, mechanical systems, and even biological samples taken on board. Therefore, it is crucial to have effective thermal control to avoid overheating or freezing, which can result in malfunctions or mission failure.

The thermal control subsystem utilizes a combination of passive and active techniques to manage the spacecraft's thermal environment.

- **Passive thermal control:** It relies on the natural properties of materials to control temperature. This involves utilizing insulation materials, reflective surfaces, and heat-absorbing or dissipating structures to passively control heat transfer. Many of the components are typically low density and lightweight and have low thermal conductivity. One example of passive thermal control is the use of radiators to disperse excess heat generated by onboard electronics, and they are coated to enhance heat dissipation. Furthermore, one can orient these radiators away from the Sun, facilitating the emission of excess heat into space. To reduce heat transfer through conduction, one can enhance this by utilizing thermal blankets composed of multiple layers of reflective and insulating materials.

- **These mechanical devices actively control heat generation and dissipation:** Heat generation and dissipation can be actively regulated by circulating fluids or gases through heat exchangers, radiators, and thermal loops. This helps transfer heat away from critical components or distribute it to areas that need more heating. A notable application of active thermal control is to raise the temperature of critical components during periods of extreme cold. There are several types of active thermal control systems, such as the following:

  - Radiators: Radiators are employed to dissipate heat generated by equipment on the spacecraft. They function by emitting heat into space through a large surface area. Radiators can be passive, such as simple metallic fins, or active, such as heat pipes or fluid systems with a pump.

  - Heaters: Heaters are employed to maintain equipment and systems warm in inhospitable environments. They can stop fluids from freezing or keep electronics operating within their temperature range. Heaters are capable of being resistive, such as wire or film heaters, or induction-based.

  - Coolers: Coolers are employed to lower the temperature of equipment and systems in hot environments. They have the ability to be thermoelectric, which utilizes a temperature gradient to generate electricity, or mechanical, such as vapor-compression refrigeration systems.

  - Phase change materials: Phase change materials are substances that absorb or release heat as they change phase, such as from solid to liquid. They are often used to stabilize the temperature of a system or to retain heat for later use. Some options are wax or paraffin-based materials.

Thermal control can also handle heat generated by onboard systems, making them more complex and challenging to oversee than their passive counterpart. Many high-power electronic components, propulsion systems, and other scientific instruments necessitate efficient heat removal to prevent overheating. This is why thermal control can also be employed to redistribute the generated heat, making the systems more complex and challenging to manage than their passive counterpart.

However, the complexity of these systems may vary depending on the spacecraft's mission, size, and specific thermal requirements. In large spacecraft or space stations, multiple subsystems with different temperature ranges may handle the thermal environment in different sections or modules. These subsystems collaborate in tandem to ensure that each component operates within its optimal temperature range, minimizing the risk of thermal stress or degradation. Yet, advancements in thermal control technologies persist to be a focus in spacecraft design and development. With the use of advanced materials, lightweight heat pipes, state-of-the-art radiators, and efficient thermal management algorithms, thermal control solutions will continue to be crucial for the success and longevity of future missions. These advancements result in enhanced performance, reduced mass, and increased mission capabilities (Figure 2.4).

Figure 2.4 A special radiation vault being placed onto NASA's Juno spacecraft to dissipate excess heat and slow the aging effect of radiation on the electronics. This demonstrates the critical importance of shielding sensitive spacecraft components from the harsh space environment, which includes both radiation and thermal challenges.

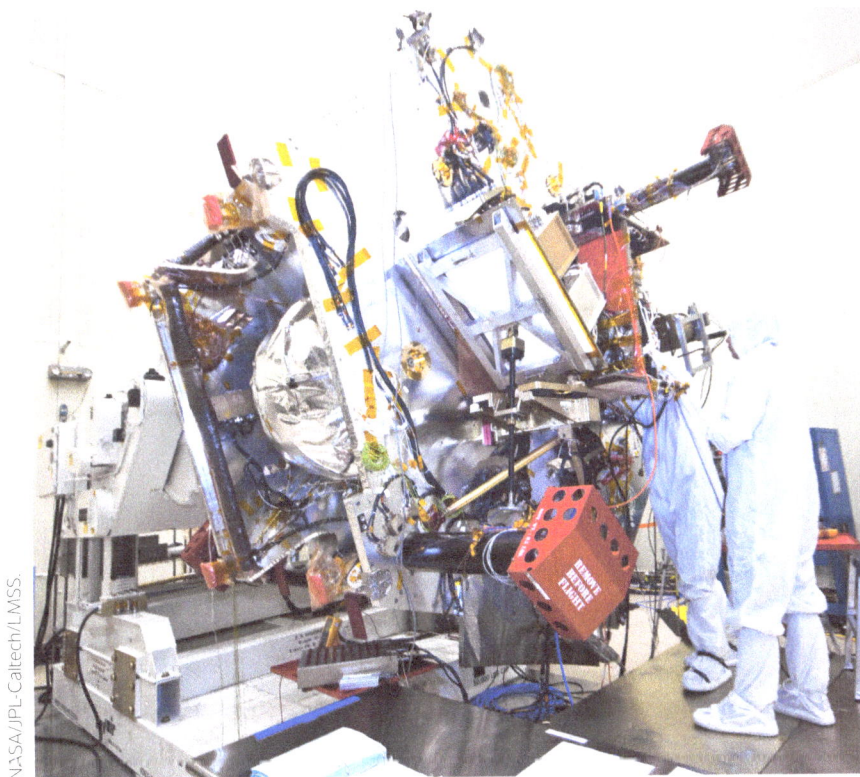

NASA/JPL-Caltech/LMSS.

2.8.
Propulsion Systems

The propulsion subsystem's main goal is to generate thrust, the force for spacecraft acceleration. Typically, the propulsion subsystem generates this thrust by emitting a high-velocity stream of exhaust gases in the opposite direction, following Newton's third law of motion: "Every action has an equal and opposite reaction." The law is the fundamental basis upon which the mechanics of pressure and force can drive the spacecraft forward. The propulsion system operates on this principle through the utilization of a combination of propellants within a combustion chamber, where they undergo a chemical reaction to generate hot gases. These are accelerated and ejected through a **nozzle** at high velocity. This ejection of gases imparts momentum to the engine, resulting in a **thrust** force. The ejection of high-velocity matter essentially causes the motor structure to go through a reaction, resulting in the force. This phenomenon can be observed in everyday scenarios, like the recoil of a fired gun or the backward push of water flowing out of a garden hose nozzle. However, comprehending the physics behind this phenomenon necessitates visually exploring the mechanics of pressure and force that occur within the combustion chamber.

As shown in Figure 2.5, a combustion chamber contains an opening, or nozzle, through which gas can escape. Within the chamber, the pressure distribution is asymmetric, with pressure fluctuating little throughout the chamber, but decreasing somewhat near the nozzle. The gas pressure exerts a greater force on the bottom of the chamber than the external pressure, leading to a resultant force caused by the pressure difference. This force causes the chamber to move upward in the opposite direction of the gas jet. To generate high speed, higher temperatures and pressures are necessary, which are achieved by utilizing fuel and minimizing the molecular weight of the exhaust gases as much as possible. It is also necessary to minimize the gas pressure as much as possible inside the nozzle by creating a large section ratio, or expansion ratio. This means that the combustion chamber of a rocket engine must endure the high pressure and temperature generated during the combustion process without degrading or failing.

Figure 2.5 Principle of propulsion.

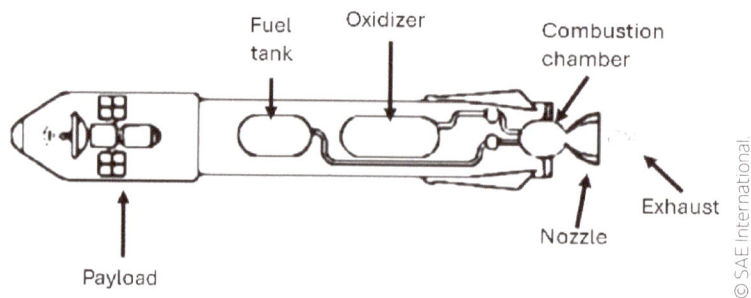

The injector supplies fuel to the oxidizer for efficient combustion. However, this injector can also perform many other functions associated with combustion stability and cooling processes. With the former, shock, detonation, and local disturbances may produce waves in the chamber, leading to fluctuations in mixing or propellant flow. This can trigger pressure oscillations with high-amplitude waves, referred to as combustion instability. This generates high levels of vibration and heat flux that can be very destructive. Therefore, obtaining stable combustion is a major focus of the design and development effort, as high performance can become secondary if the injector is easily triggered into destructive instability.

2.9.
Structural Integrity

The structure ensures support, stability, and protection for the payload and other subsystems throughout the mission. It serves as the physical framework of the spacecraft, guaranteeing structural integrity in the demanding conditions of space. The primary goal is to withstand the extreme environments faced during launch, in orbit, and during reentry, as well as the various forces acting on the spacecraft. The structure must be able to tolerate the intense vibrations, acceleration forces, and thermal fluctuations experienced during launch, as well as the vacuum of space, micrometeoroid effects, and temperature extremes. As a result, engineers thoughtfully develop the structure to meet the specific requirements of the type of mission. The structure must meet certain prerequisites, such as being lightweight, robust, and made of high-strength materials like alloys, composites, or specialized materials for durability in space

The primary framework is crucial for supporting key spacecraft components like the payload, propulsion system, and subsystems. It must exhibit enough rigidity to maintain the spacecraft's shape and stability, ensuring proper alignment and operation of the various systems. To optimize structural integrity and load distribution, secondary structures, such as struts, braces, and supports, are often incorporated. These elements assist in the distribution of the loads and stress throughout the spacecraft, preventing excessive bending or deformation. The asset's outer shell will act as a barrier, protecting and shielding the spacecraft from the extreme temperatures of space and atmospheric reentry. Therefore, it is essential to design it with thermal control so that it can dissipate heat generated. Besides these structural elements, the spacecraft may also include deployable structures, such as solar panels, antennas, or instrument booms. These structures extend or expand after launch to provide the functionality for the mission.

Therefore, the design of spacecraft must consider factors, such as structural constraints, launch vehicle limitations, mission requirements, and operational considerations. Many structural engineers will use advanced analysis techniques, including finite element analysis (FEA) and computer-aided design, to ensure the structural integrity and performance of the spacecraft. Ongoing research development currently concentrates on the adoption of new materials, construction methods, and innovative structural concepts to enhance spacecraft performance. Efforts are also underway to use lightweight yet strong materials, such as carbon fiber composites, as well as additive manufacturing techniques for forming complex structures. Furthermore, deployable or inflatable structures are being investigated to reduce launch mass and increase mission capabilities.

The ISS, with its approximate length of 109 m and weight exceeding 420,000 kg, is a remarkable example of structure integrity in space history. Its construction started in 1998 and has involved many assembly missions using space shuttles, Russian Soyuz spacecraft, and robotic systems. Putting it together in orbit was an intricate process that demanded meticulous engineering and thoughtful planning, and ensured the structural integrity of various interconnected modules and trusses. These modules and trusses provide support and living space for astronauts. These modules withstand the demanding conditions of space, such as micro-gravity, extreme temperature fluctuations, and micrometeoroid impact. The structure must also withstand the pressures experienced by the dynamic loads imposed by the crew's activities and experiments conducted on board.

Throughout its operational history, the ISS showed exceptional structural integrity. It has endured various space debris effects, such as small fragments and even larger objects, without compromising the safety of the crew or the station's functionality. The ISS's robust construction proved its ability to withstand the harsh conditions of space and maintain its structural integrity over long-duration missions. The knowledge gained from the design, construction, and operation of the ISS has significantly influenced the development of future space structures and habitats (Figure 2.6).

Figure 2.6 Teams make arrangements for the completed Orion pressure vessel for the Artemis IV mission, emphasizing the importance of structural integrity in the Orion crew module's primary structure. All other elements of a spacecraft come together upon the structural components, forming the core. The structure needs to withstand launch stresses and harsh space conditions.

NASA/Michael DeMocker.

2.10. Summary

This chapter summarizes spacecraft systems. It brought attention to some of the ongoing advancements in technology that are crucial for meeting mission requirements and ensuring future mission success. A key factor in spacecraft design is the delicate balance between cost, performance, and safety requirements. This task transforms into a challenging one when many subsystems might have conflicting priorities, requiring careful evaluation and trade-offs.

Therefore, a comprehensive understanding of the mission goals and objectives is essential to ensure that the spacecraft system design matches the intended specifications. The chapter stresses the primary significance of maintaining the reliability of these systems. The design and operation of spacecraft must carefully tackle the host of unpredictable challenges presented by the unique and unforgiving environment of space. Rigorous planning and preparation are vital to withstand the demanding conditions of spaceflight and achieve the intended mission objectives. By highlighting the importance of continuous technological advancements, the chapter acknowledges the dynamic nature of spacecraft design and the ongoing pursuit of innovation.

References

2.1. Wertz, J.R. and Larson, W.J. (Eds), *Space Mission Analysis and Design* (Torrance, CA: Microcosm Press and Kluwer Academic Publishers, 1999).

2.2. Fortescue, P., Swinerd, G., and Stark, J. (Eds), *Spacecraft Systems Engineering*, 4th ed. (Chichester: Wiley, 2011).

2.3. Sellers, J.J., Astore, W.J., Giffen, R.B., and Larson, W.J., *Understanding Space: An Introduction to Astronautics*, 3rd ed. (New York: McGraw-Hill, 2005).

2.4. National Aeronautics and Space Administration, "NASA Systems Engineering Handbook (NASA/SP-2007-6105 Rev1)," Washington, DC, 2007.

2.5. Braeunig, R.A., *Rocket and Spacecraft Propulsion: Principles, Practice, and New Developments* (Chichester: Springer, 2013).

2.6. Griffin, M.D. and French, J.R., *Space Vehicle Design*, AIAA Education Series, 2nd ed. (Reston, VA: American Institute of Aeronautics and Astronautics (AIAA), 2004).

2.7. Pardini, C. and Anselmo, L., "Collision Risk of Spacecraft and Space Debris in Low Earth Orbit," *Advances in Space Research* 45, no. 6 (2010): 709-716.

How can we keep a spacecraft reliable during its long journey in space?

This question is at the heart of ensuring the success of missions and highlights the need for effective health management strategies for space vehicles. The primary goal of health management in space is to prevent system failures and extend the spacecraft's life. This is done by spotting and fixing potential issues before they escalate into significant concerns. To achieve this, continuous monitoring and analysis of the vehicle's data and systems are essential. We need to keep a close eye on various sensors and instruments onboard, like temperature, pressure, and vibration sensors. By analyzing these data with special algorithms, we can identify any unusual patterns or issues. Once we detect a problem, we can take corrective actions. This might involve adjusting how the spacecraft operates or repairing its systems or components. Long or high-risk missions, like planetary exploration or satellite operations, require crucial health management. By monitoring and managing the health of its systems and components, we ensure that the mission meets its safety and reliability requirements. This also helps in reducing the risk of system failures and the potential loss of valuable scientific data.

Reliability engineering principles have long underpinned the development of spacecraft systems, highlighting the early recognition of failure prevention strategies in long-term missions [3.1]. This chapter will highlight the significance of health management systems and provide examples of notable failures that could have been avoided. This approach helps us understand how crucial it is to maintain the health of a spacecraft for the success of space missions.

3.1.
Successful Space Missions

Health management systems are key to the smooth running of various vehicles and systems, playing a big role in ensuring safety, cost-effectiveness, sustainability, and efficiency. These systems work by monitoring the health of a vehicle or system, identifying any faults, and suggesting fixes.

Different areas use them, such as aerospace (e.g., Boeing 777 and Mars rovers), cars, and nuclear powerplants. The level of complexity and automation can greatly differ depending on its purpose. Understanding how a system might fail is crucial for setting up any monitoring technology. This involves analyzing data from sensors such as those for temperature, pressure, and vibration. Advanced data analytics help spot any unusual patterns. The system uses algorithms to predict how it will behave under various conditions.

Learning from past tragedies like the Apollo, Challenger, and Columbia accidents, thorough health checks of vehicles are vital. These checks are essential for quickly determining crew safety and mission changes. NASA's experience with the Space Shuttle showed the high cost of operating such vehicles. Future space missions need to be more sustainable. Monitoring systems are being designed to constantly check the health to meet these future needs. They can diagnose problems and give quick advice on what to do

next. These systems need to work under different mission conditions to keep the vehicle safe and running well. This requires integrating advanced sensors, data analytics, and ML. Ensuring these systems work correctly (verification and validation) is important for their reliability and effectiveness.

These advanced systems can benefit space missions. They can quickly find and identify issues, which are especially useful for missions that need fast responses. For longer missions, automated fault finding and predictions can make the vehicle more self-reliant and cut down on costs by reducing the need for constant ground crew monitoring. Good health data help with preflight checks, mission planning, maintenance, and safety. An effective system should help the crew focus more on their mission goals, increasing the chances of success. In-flight maintenance and advanced predictions can lower the need for repairs and overall costs in reusable systems. Being able to address challenges, manage failures, reduce faults, and perform maintenance based on the vehicle's condition can increase the duration and sustainability of space missions.

Let us look at some successful missions that highlight the importance of health monitoring in long and complex space missions.

- Mars Curiosity rover: Launched by NASA in 2011, the Mars Curiosity rover used a sophisticated monitoring system to detect any potential issues during its long mission. It used advanced fault detection and isolation algorithms to identify and correct any potential problems, allowing the rover to continue its mission without interruption.

- The ISS: It has maintained continuous manned operations for over 20 years, relying on a complex safety system for crew well-being. This includes sensors and instrumentation to monitor the health of the station's systems, as well as SW and data analysis tools to detect and diagnose any potential issues.

- HST: The HST has been in operation for over 30 years and relies on a sophisticated system to maintain its continued success. The health management system includes advanced data analysis tools and fault detection algorithms to detect and diagnose any potential issues with the telescope's systems.

- Hayabusa2: Launched by the Japanese space agency JAXA in 2014, Hayabusa2 used a sophisticated system to ensure the success of its mission to study the asteroid Ryugu. It featured sophisticated fault detection and isolation algorithms, along with real-time monitoring and data analysis tools to ensure the health and safety of the spacecraft.

- The SpaceX Dragon: Spacecraft that transports cargo and crew to and from the ISS using a sophisticated system to ensure mission success. It includes sophisticated fault detection and isolation algorithms, in addition to real-time monitoring and data analysis tools to ensure the well-being and security of the spacecraft and its crew.

Recent advancements in technology have allowed for some features to be implemented onboard as an essential factor in ensuring the success of space missions.

3.2.
The Signs of Premature Failure

We have already seen how technology can predict when equipment might fail. But it is important to remember that failures do not always happen suddenly or without warning. By using models that recognize how equipment behaves, we can identify which parts might fail early and replace them before they do. Usually, designers create these diagnostic and prognostic algorithms for specific types of equipment and applications. So, while they can theoretically detect early signs of wear or failure, they are not one-size-fits-all solutions. Sometimes, even if the signs are there, the technology may not always accurately predict a failure. To improve this, we focus on using specialized algorithms to detect early warning signs. *NASA's Probabilistic Risk Assessment Guide* outlines methods for identifying risks and ensuring that potential failures are addressed before they escalate [3.2]. Training personnel to recognize these signs can also help. By linking different behaviors to when a failure is likely to happen, this technology can help us avoid unexpected problems and keep things running smoothly. By understanding how long equipment can last before it fails, we can better judge its reliability. For example, if we see certain patterns in the operational data, we could predict that a part will fail within a year.

Spacecraft already widely uses telemetry, which is data about how equipment is performing. Algorithms can use this data to figure out how reliable equipment is, whether it is going to last for hours, days, weeks, or months. This not only saves money but also improves reliability. However, various factors, from vibrations to external influences like neutrinos, can sometimes cause these warning signs to be missed or misinterpreted. We need to spot these unusual patterns, provide data for fixing problems, and work with minimal human input for large systems to reduce overall costs. Successfully doing this can reduce the need for constant ground control and monitoring.

In the past, missions were expensive and complex, with limited observation of the ground. Without quick and accurate fault diagnosis and fixing, keeping missions within budget and successful was tough. This is where health monitoring can make a difference. It uses sensors, algorithms, and diagnostic tools to find and fix potential problems before they get serious. By constantly checking things like propulsion, power supply, communication, and payload, we can spot any unusual patterns or signs of failure. When it finds a problem, it can trigger the right response, like sending alerts, fixing the issue, or switching to backup systems. This approach reduces the risk of system failure, prevents damage, and makes them last longer.

In the aviation industry, health monitoring systems (HMS) are now a key part of making sure assets are ready to fly and managing their various parts and systems effectively. This technology allows aircraft to check their condition and deal with many problems on their own. However, there is a challenge: early signs of wear or potential failure hide in data that seems normal from equipment that appears to be working fine. This can lead to the belief that there is no need for further checks or analysis. This leads to failing devices passing factory tests without detection. The problem is that these parts can then fail within their first year of being used in actual flight conditions. This shows a gap in how we detect and handle potential failures in aviation equipment, highlighting the need for more thorough and sensitive HMSs.

- Early signs of aging or failure often resemble other common behaviors found in data, such as signal noise, noisy data, equipment cycling transients, and sensor failure. It requires specialized training for personnel to differentiate between the early signs of aging or failure and these other common behaviors.

- The presence of early signs of aging or breakdowns cannot be used to compare or identify them with previous equipment or malfunctions because of minor variations within each part. The uniqueness of each failure event prevents the replication of patterns.

- Reliability analysis relies on the assumption that equipment failures are random and instantaneous, known as the Markov property. Therefore, it is not possible to link any behavior in the test data before a failure to the failure itself. This means that previous behavior in the test data cannot be linked to equipment failures, making them unpredictable and unpreventable.

In contrast, certain behavior refers to where the outcome is 100% predictable. Different names exist in various industries, such as failure precursors, transients, predictive markers, and prognostic identifiers. In the commercial and military aircraft industry, maintenance personnel may refer to such deterministic behavior as "Cannot Duplicate" (CND), "No Failure Found" (NFF), etc. There are many other terms for it too, which can be confusing.

Although this book does not mainly focus on deterministic behavior, it is worth mentioning because it is a significant issue in aircraft maintenance. When maintenance teams encounter a problem that appears once but cannot be found or replicated during testing, it becomes a significant challenge. Elusive issues like this can consume a large part of an airline's maintenance budget. Teams spend a lot of time and resources trying to track and fix these unpredictable problems.

3.3.
Learning from the Past

So, we know there are some initial signs that we can use. Early indications show that real-time monitoring and analysis could have prevented or lessened many failures. Let us look at some examples:

- Space Shuttle Challenger (1986): The failure of the Space Shuttle Challenger resulted in the loss of the entire crew. Post-investigations suggested that, by monitoring the O-ring, any degradation could have been detected, providing an early warning to prevent the launch and avoid a catastrophic failure.

- Mars Climate Orbiter (1999): The Mars Climate Orbiter mission failed because of a navigation error caused by conflicting units of measurement. The spacecraft's SW used metric units, while the ground-based systems used imperial units. Mission control could have detected this discrepancy and been alerted to prevent the incorrect trajectory and subsequent loss of the spacecraft.

- Genesis (2004): The Genesis mission aimed to collect samples from the solar wind for scientific analysis. However, during reentry, the parachute system failed to deploy, causing the spacecraft to crash upon landing. Continuous health monitoring of the parachute system for any faults or anomalies could have enabled timely maintenance (or repairs) to ensure a successful landing.

- SpaceX CRS-7 (2015): It aimed to resupply the ISS but failed shortly after liftoff because of a structural failure in the second stage of the Falcon 9 rocket. Detecting the abnormal behavior or stress on the rocket structure could have prevented this. They could have provided an early warning to abort the mission and prevent the loss of the spacecraft and payload.

- Soyuz 11 (1971): The Soyuz 11 mission ended tragically when a cabin vent valve improperly opened during reentry, causing the crew to asphyxiate. Continuous monitoring of the cabin environment could have detected abnormal pressures (or gas composition) and triggered appropriate actions to maintain a safe environment for the astronauts.

These examples highlight the importance of SHM and how it could have potentially prevented or mitigated catastrophic failures during spacecraft missions. Continuously monitoring critical systems is a key aspect here, for improving safety, reliability, and mission success [3.1].

3.3.1.
The Challenger Failure
In 1986, the Challenger's Solid Morton-Thiokol Rocket Motor (SRM) engineering manager warned both Morton-Thiokol and NASA Space Shuttle management about the potential failure of the O-rings. This prediction was based on earlier O-ring failures that had occurred at warmer outside temperatures than expected for the Challenger launch. The outside temperature was colder on this occasion. The Morton-Thiokol engineering manager knew the Challenger's O-rings would be more rigid, increasing the risk of fuel leakage. The early signs of premature failure present on the retrieved Challenger SRM

O-rings confirmed these failure predictions, as both O-rings failed. Despite this, the warnings were ignored as it was believed that equipment failures are instantaneous and random, making them unpredictable.

However, the real-time telemetry and video data during the launch in 1986 identified the failure of the rubber O-rings at liftoff. The video footage captured the burn-through of the solid booster O-rings at ignition, a critical moment of the launch. This burn-through could have been identified by experts skilled in prognostic analysis had they been evaluating real-time telemetry. They would have instructed the Challenger flight team to terminate the launch and douse the solid rocket motors with water to prevent further damage. The telemetry data have also shown a significant drop in solid rocket motor fuel tank pressure after the hole in the side of the solid motor tank occurred. It is worth mentioning that there was no subsystem personnel stationed at the strip chart recorders in the Space Shuttle mission control room. They were not evaluating the Challenger's telemetry before liftoff. This absence of engineering analysts, who could have searched for predictive markers and identifiers in the Shuttle telemetry data, was primarily driven by cost-saving measures. Again, the agency's management believed that failures were instantaneous and random, adhering to the Markov property, which led them to conclude that failures could not be predicted or prevented! The oversight of crucial information was contributed by the agency's management deeming the presence of engineering personnel unnecessary.

Despite recording and displaying Space Shuttle telemetry data and having an analysis for system malfunctions, the team ignored its significance. Their organizational culture and perception led to the failure to utilize recorded data. They believed that failures were random and unpredictable, which resulted in the lack of priority given to analyzing telemetry for prognostic identifiers. In terms of emergency actions, they could have taken some measures if they had detected a failure in the solid rocket motor through real-time telemetry and video observations. This action involved igniting the Space Shuttle's primary engine early while in motion, using fuel from the central tank, followed by early jettisoning of the solid rocket motors. However, because of the absence of engineering personnel dedicated to evaluating telemetry in real time, nobody could identify the need for a launch abort. Omitting this personnel, who could have recognized the predictive markers and made informed decisions, was a contributing factor to failing to prevent the disaster.

It is worth noting the similarity between the Space Shuttle Challenger failure and the subsequent failure of the Space Shuttle Columbia (discussed next) (Figure 3.1). In both cases, personnel consciously chose to ignore available information provided by real-time telemetry. Previous launches had experienced no problems during these phases, leading management to believe that the technical team was unnecessary. Proper utilization of telemetry data can indeed help identify predictive markers, prevent failures, and ensure the safety and success of future space vehicle programs.

Figure 3.1 Liftoff of Shuttle Challenger for STS 51-L mission on January 28, 1986, with a crew of seven astronauts and the Tracking and Data Relay Satellite. An accident 73 sec after liftoff claimed both crew and vehicle.

NASA.

Lessons to learn:

- Even if the equipment appears to be functioning normally, one should never ignore early signs of equipment failure. It is important to take initiative and thoroughly investigate and analyze.

- Expert personnel should receive special training to discern the early signs of failure from commonly occurring behavior in data associated with signal noise, equipment transients, and sensor failures.

- We cannot assume that equipment failures happen instantaneously and randomly. Prognostic analysis and real-time telemetry evaluation can help identify failure precursors and prevent catastrophic incidents.

- We should diligently monitor real-time telemetry data for predictive markers and identifiers. Make sure to diligently monitor strip chart recorders and other data collection methods as they provide important identifiers.

- Believing that failures are unpredictable and cannot be prevented can lead to complacency and oversight.

- It is crucial to recognize the value of information provided by telemetry and make informed decisions based on those data.

- Properly trained engineering personnel should be present during launch readiness and operations to evaluate real-time telemetry and identify potential failures.

- Cost-saving measures should not compromise safety and reliability.

3.3.2.
Columbia Failure

On February 1, 2003, Space Shuttle Columbia disintegrated upon reentry, killing all seven astronauts (Figure 3.2). The disaster was due to damage from debris striking the wing during launch, which was not thoroughly investigated during the mission. During launch, debris struck Columbia's wing, causing damage. This was noticed by NASA's Intercenter Photo Working Group on the second day but was not clearly visible in any launch videos [3.3]. As a result, requests for on-orbit images to assess the damage were canceled by mission management, believing it was not necessary. Engineers from NASA, United Space Alliance, and Boeing formed a Debris Assessment Team to analyze the damage. The models suggested significant damage, but experiences led the team to dismiss these findings. Also, multiple requests for imaging of the orbiter were also canceled, partly because of concerns about interrupting science operations. The mission management team had downplayed the risk, citing past foam strike incidents that did not result in any disasters. As Columbia reentered the atmosphere, the damage

allowed hot air to penetrate and melt the wing's aluminum structure, leading to the orbiter breaking apart. The crew experienced rapid decompression and were likely unconscious or deceased before the breakup completed.

This emphasized the need for thorough evaluation of telemetry and real-time data to detect and address damage early. The disaster also highlighted flaws in NASA's organizational culture, similar to those before the Challenger disaster, and recommended improvements in communication, risk assessment, and safety programs.

Lessons to learn:

- We should not overlook or disregard real-time telemetry and data, especially when it comes to safety-critical systems. Neglecting to evaluate information can have severe consequences.

- Prognostic analysis and trained personnel are crucial for assessing data and identifying potential failures. Taking a proactive approach and considering all available information can help prevent disasters.

- Telemetry can provide valuable insights into the condition of spacecraft components. Analyzing these data is essential to detect damage or anomalies.

- Ignoring information because of time-consuming analysis or perceived low risk is a dangerous practice. It is important to allocate resources and prioritize thorough diagnostics to ensure a comprehensive understanding of the situation.

- Decision-making processes should thoroughly consider identifiers such as previous problems, real-time video data, and internal telemetry.

Figure 3.2 Space Shuttle Columbia on the launchpad for STS-107: (a) The circled area on the external tank is the left bipod foam ramp, which would fall off during launch and the location that was damaged by the foam is circled in accordance with the image in the Columbia Accident Investigation Board report. (b) The right image shows a close-up of the left bipod foam ramp that broke and damaged the orbiter wing.

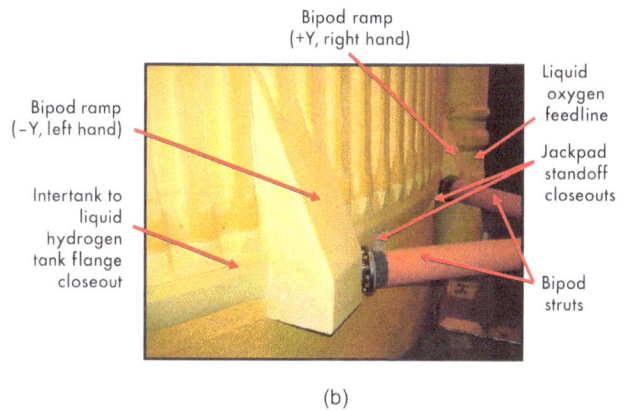

(a)

(b)

- Embracing predictive technology and analysis can help prevent accidents by ensuring that all available data are considered and used for decision-making.

- Personnel without adequate data reduction and analysis experience should not be solely responsible for decision-making. Decision-makers should accumulate, analyze, and make all relevant data available.

- We should recognize and apply the value of telemetry and the need for thorough analysis to future projects.

- The commitment to using prognostic technology across all future NASA missions, manned and unmanned, is vital for reducing the risk of catastrophic failures.

NASA's exploration technology program review underscores the importance of advancing technology in both manned and unmanned missions, particularly for risk mitigation and SHM [3.4]. Analyzing failures and near misses provides invaluable lessons for improving human spaceflight, especially regarding SHM and risk assessment [3.5].

3.3.3.
The Mars Climate Orbiter Mission

Launched in 1998 by NASA, this vehicle was designed to study the Martian climate, atmosphere, and surface conditions. However, on September 23, 1999, as the spacecraft was supposed to enter Mars' orbit, contact with the orbiter was lost, and it ultimately disintegrated in the planet's atmosphere. The mission failure was attributed to a critical navigation error caused by conflicting units of measurement used in the spacecraft and the ground-based systems. The onboard SW used metric units (specifically, newton-seconds), while the ground-based systems employed imperial units (pound-seconds) to calculate and relay navigation data. This deviation went unnoticed during the mission planning and testing phase, leading to the spacecraft being on a path that brought it too close to Mars (Figure 3.3). The resulting atmospheric forces caused the spacecraft to experience excessive aerodynamic stresses, ultimately leading to its destruction.

The error could have been prevented by employing a combination of sensors, data analysis algorithms, and predictive models to monitor the health and performance of critical systems in real time. In particular, the system would have included sensors to monitor various parameters related to the spacecraft's navigation and propulsion, including acceleration, position, velocity, and path, and integrated SW algorithms to process and analyze the collected data.

One of the key features is its ability to identify discrepancies and anomalies in the data. Continuously comparing the spacecraft's navigation data with predicted values can help identify inconsistencies. This would have triggered an alert, indicating a potential navigation error. Furthermore, ML algorithms can be used to recognize patterns and trends in the information. By training the system with historical mission data and known successful trajectories, it could have learned to identify abnormal or unexpected deviations in the navigation parameters. This algorithm could detect conflicting units of measurement and raise a warning, emphasizing the need for immediate attention and resolution. To prevent the navigation error from occurring, a possible solution would have been to integrate a standardized unit conversion module within the spacecraft's SW and ground-based systems. This module would have ensured standardized measurement units throughout the mission, eliminating the potential for unit conversion errors. Additionally, a validation/verification component is required to cross-check the navigation data received from different sources.

Lessons to learn:

- It is crucial to ensure consistent units of measurement throughout the mission, especially when dealing with critical navigation and trajectory data.

- Thorough verification and validation processes should be implemented to detect and address any inconsistencies or errors in data inputs.

- Effective integration between different systems and components is essential to ensure accurate and reliable operation.

- Clear and effective communication between teams involved in different stages of the mission is vital to prevent misunderstandings and errors.

- Rigorous quality control procedures should be in place to identify and rectify any issues or anomalies before launch.

Figure 3.3 The Mars Climate Orbiter was supposed to monitor Mars' atmosphere and surface for two Earth years. This included tracking atmospheric dust, water vapor, and seasonal changes, providing insights into Mars' climate history and potential liquid water reserves beneath the surface. The mission ended in failure on September 23, 1999. The spacecraft was lost due to a navigation error caused by a mix-up between metric and imperial units.

NASA.

- Comprehensive risk management strategies, including proactive identification and mitigation of potential risks, are crucial for mission success.

- Continuous monitoring of critical parameters and system health is essential to identify deviations or anomalies early on.

- Designing spacecraft systems with built-in fault tolerance mechanisms can help prevent mission failures in the event of unexpected issues.

- Prompt and preemptive problem-solving approaches should be employed when anomalies or deviations are detected, ensuring swift corrective actions.

- Establishing a robust lesson-learned process to analyze failures and implement improvements for future missions is essential for continuous progress and success.

3.3.4.
Genesis Mission

This was launched in 2001 by NASA, aimed at collecting samples of the solar wind and returning them to Earth for in-depth analysis. However, on September 8, 2004, the mission encountered a critical failure during the capsule's reentry and landing, leading to a crash in the Utah desert. This incident started a comprehensive investigation to determine the causes of the failure and identify valuable lessons for future missions.

The primary factor contributing to the Genesis mission failure was the malfunction of the parachute system designed to facilitate a safe landing for the sample return capsule. The investigation revealed that the parachute deployment sequence was disrupted due to the incorrect installation of accelerometers. These sensors were inadvertently installed in an

inverted orientation, leading to erroneous altitude measurements and premature deployment of the parachutes.

Consequently, the parachutes failed to slow down the descent adequately, resulting in a hard landing and severe damage to the capsule. The misalignment of the accelerometer sensors was attributed to a lack of thorough inspection and quality control measures during the spacecraft assembly process. The oversight in ensuring the correct installation of critical components highlighted the significance of paying attention to detail in the manufacturing process. Failing to detect and rectify this error before launch ultimately led to the mission's unsuccessful landing and the subsequent loss of scientific data.

Incorporating redundancy and robust contingency planning in critical systems was underscored by the failure of the Genesis mission (Figure 3.4). Had the parachute system been equipped with redundant sensors or backup mechanisms, the mishap could have been mitigated. The incident prompted a renewed focus on designing fail-safe systems that can compensate for single-point failures, ensuring mission success even in the face of unexpected challenges.

Lessons to learn:

- The mission highlighted the critical importance of rigorous quality control measures throughout the spacecraft assembly process. It emphasized the need for meticulous inspection and verification of all components to ensure correct installation and functionality.

- The Genesis failure underscored the significance of redundancy in critical systems and the need for thorough failure analysis. It emphasized the significance of conducting

Figure 3.4 The Genesis spacecraft in the Payload Hazardous Servicing Facility, with both solar arrays deployed. It was designed to collect solar wind particles and return these samples to Earth for scientists to study the Sun's composition and the solar system's origin. The white object at the front houses the collector arrays.

NASA

comprehensive testing and analysis to identify potential failure modes and implement appropriate countermeasures.

- The mission failure demonstrated the importance of clear and effective communication among the various teams involved in spacecraft development, including design, manufacturing, and testing. It emphasized the need for seamless collaboration to identify and address potential issues.

- The Genesis failure prompted a reevaluation of spacecraft manufacturing processes and the implementation of improved quality control measures. It emphasized the need for a culture of continuous improvement, where lessons learned from failures are used to enhance future missions.

- The mission emphasized the significance of paying meticulous attention to detail at every stage of spacecraft development. It highlighted the potential impact of seemingly minor errors or oversights on mission success.

3.3.5.
SpaceX CRS-7

This mission, launched on June 28, 2015, was a cargo resupply mission to the ISS, conducted by SpaceX, a private aerospace company. However, just over two minutes after liftoff, the mission experienced a catastrophic failure, resulting in the loss of the Falcon 9 rocket and the Dragon spacecraft. This event marked a significant setback for SpaceX and highlighted the complexities and challenges associated with space exploration and rocketry.

The SpaceX CRS-7 mission failure was the result of several factors. First, the investigation revealed that a critical strut designed to secure a helium pressure vessel within the Falcon 9 rocket's second stage failed under high forces during launch. The strut's inadequate strength and reliability became apparent as it failed to withstand the stresses encountered during the mission. Second, the failure was attributed to manufacturing oversight, with insufficient testing and evaluation of the strut's performance and reliability. This oversight exposed a gap in the quality control processes, emphasizing the importance of thorough testing and stringent manufacturing standards in aerospace engineering. The mission was lost due to the failure of the strut, leading to the disintegration of the second stage. This incident demonstrated the cascading effects that a single component failure can have on the entire system, emphasizing the critical nature of each component's reliability and robustness.

Lessons to learn:

- The failure underscored the need for stringent quality assurance processes in the design, manufacturing, and testing of spacecraft components. Thorough evaluation, testing, and verification at every stage of production are vital to identify and address potential weaknesses and failures.

- The incident highlighted the significance of redundant systems and robust design in spacecraft engineering. Implementing redundant components and ensuring their reliability can mitigate the impact of a single component failure and increase mission success rates.

- The failure prompted SpaceX to review and enhance its design, manufacturing, and quality control processes. This commitment to continuous improvement and learning from failures is crucial for the advancement and success of space exploration endeavors.

3.3.6.
Soyuz 11 Failure

The mission, launched on June 6, 1971, was intended to be a ground-breaking achievement in space exploration. However, it ended in a tragic failure that resulted in the loss of three cosmonauts. The incident highlighted the critical importance of crew safety and the need for improved spacecraft design and emergency protocols.

Soyuz 11 was the first manned mission to the Salyut 1. The three cosmonauts aboard, Georgi Dobrovolski, Viktor Patsayev, and Vladislav Volkov, had spent almost three weeks conducting experiments and research. Their planned return to Earth was eagerly anticipated, but tragedy struck during reentry. As the Soyuz 11 spacecraft prepared for reentry, a faulty valve in the separation mechanism caused the cabin to vent its atmosphere into the vacuum of space. Upon touchdown, recovery teams discovered that all three cosmonauts had perished tragically due to asphyxiation.

The investigation into the Soyuz 11 failure identified two primary causes. First, a ventilation valve failed to close properly, leading to the loss of the cabin atmosphere. Second, the design of the spacecraft's emergency procedures did not account for such an event, as the crew was not equipped with individual emergency suits.

The Soyuz 11 failure had a profound impact on the space industry. It served as a wake-up call, highlighting the need for robust emergency systems and enhanced crew safety measures. As a direct response to the incident, contingency plans were implemented in subsequent Soyuz missions, ensuring the crew's ability to survive a similar event. Additionally, the tragedy prompted a renewed focus on spacecraft design and safety protocols, leading to significant advancements in emergency procedures and system redundancies. International space programs, including the USSR/Russian space program, have faced similar challenges, as detailed in human spaceflight accident studies [3.6].

Lessons to learn:

- The tragic loss of the crew highlighted the paramount importance of crew safety in space missions. It emphasized the need for comprehensive safety protocols, including the provision of individual emergency suits, to protect astronauts in the event of unforeseen emergencies.

- The failure of the ventilation valve underscored the criticality of reliable life support systems. This incident prompted a reevaluation of spacecraft design to ensure robustness and redundancy in ventilation and life support mechanisms, minimizing the risk of catastrophic failures.

- The event highlighted the necessity of thorough emergency preparedness and crew training. Astronauts need to be prepared to handle unexpected situations, like the loss of cabin atmosphere, and be able to carry out emergency procedures quickly and efficiently.

- The Soyuz 11 failure catalyzed advancements in design and engineering. It prompted

rigorous evaluations of systems, components, and procedures to identify vulnerabilities and areas for improvement, resulting in enhanced safety measures and increased robustness of subsequent spacecraft.

- The event highlighted the significance of collaboration and information sharing within the space industry. Space programs and organizations globally shared lessons learned from the failure, facilitating improvements in safety standards and procedures.

- The incident highlighted the need for ongoing monitoring and evaluation of spacecraft systems throughout the entire mission duration. Regular inspections, functional checks, and maintenance procedures are crucial to identify potential issues and mitigate risks that could compromise crew safety.

- The psychological toll that space missions can have on astronauts was also recognized. This tragic event prompted a greater focus on providing adequate psychological support and counseling services for astronauts, ensuring their mental well-being during and after space missions.

3.4. Summary

One cannot overstate the importance of health management. This chapter discusses the signs of premature failure, past failures, and the valuable insights gained from them. The signs serve as crucial indicators that help identify potential problems before they escalate into critical failures. By monitoring and analyzing parameters and markers, it is possible to detect instances of interest and take proactive measures to mitigate risks. Early detection of these signs allows for timely interventions, preventing failures and enhancing overall system reliability. This chapter on examining past mission failures has provided valuable insights into the consequences of overlooking system health issues. In particular, the failures of the Challenger and Columbia disasters have been catastrophic. These (and other) incidents have highlighted the need for robust monitoring systems, comprehensive maintenance protocols, and continuous assessment of spacecraft health throughout their operational lifetimes. Moreover, the understanding obtained from these failures has paved the way for significant improvements in spacecraft.

Since then, space agencies and industry stakeholders have adopted stringent procedures for design, testing, and maintenance, incorporating redundancy, fault tolerance, and enhanced monitoring capabilities. By prioritizing these capabilities, we can significantly enhance the success rate and longevity of missions. Anticipating and preventing failures ensures astronaut safety and protects substantial investments during space exploration and research assets. This is also needed for commercializing the industry and achieving objectives like tourism, experimentation, and satellite deployment.

References

3.1.   Saleh, J.H. and Marais, K., "Highlights from the Early (and Pre-) History of Reliability Engineering," *Reliability Engineering & System Safety* 91, no. 2 (2006): 249-256, doi:https://doi.org/10.1016/j.ress.2005.01.003.

3.2.   Stamatelatos, M. and Dezfuli, H., *Probabilistic Risk Assessment Procedures Guide for NASA Managers and Practitioners*, 2nd ed. (Washington, DC: National Aeronautics and Space Administration, 2011).

3.3.   National Aeronautics and Space Administration, *Columbia Accident Investigation Board Report* (Washington, DC: National Aeronautics and Space Administration, 2003).

3.4.   National Research Council, *A Constrained Space Exploration Technology Program: A Review of NASA's Exploration Technology Development Program* (Washington, DC: The National Academies Press, 2011), doi:https://doi.org/10.17226/13054.

3.5.   Barr, S., *Evaluating Failures and near Misses in Human Spaceflight History for Lessons for Future Human Spaceflight* (Houston: The Aerospace Corporation, 2010).

3.6.   Clark, J.B., "Human Spaceflight Accidents: The USSR/Russian Space Program," in Young, L.R. and Sutton, J.P. (eds), *Handbook of Bioastronautics* (Cham: Springer, 2021), 781-796, doi:https://doi.org/10.1007/978-3-319-12191-8_65.

Good engineering practices lead to safer and more sustainable technological advancements. These practices help in making smart decisions, designing ethically, using resources wisely, and coming up with innovative solutions. To gain this knowledge, we analyze and apply data. We gather data, spot patterns, and use these insights to decide. Over time, applying these decisions teaches us valuable lessons. In SHM, there is a mix of ideas from systems engineering, reliability engineering, and data analytics. This combination helps us understand, predict, and manage issues in complex systems.

Here is how it usually works:

We take raw data from sensors and turn them into useful information.

We look for patterns in the collected data to understand the system's behavior and spot potential problems.

By analyzing the information and learning from experience, we get a deeper understanding of the system. This helps us see how it reacts to different situations and how it affects costs.

Many engineering areas are now using this approach to improve maintenance and control costs.

Let us examine how different concepts collaborate to improve engineering system reliability in this chapter.

FTA: This is a top-down, deductive analytical mode used to study the cause and effects of a specific fault. It is a crucial tool for understanding potential vulnerabilities and designing strategies to mitigate them.

- Diagnostics: Diagnostics are fundamental to health monitoring. They help in identifying the current state of the system, any anomalies, and their causes. A robust diagnostic process is the foundation for any effective system.

- Prognostics: While diagnostics deal with the present, prognostics look to the future. They predict the future state of a system, allowing for timely interventions and maintenance actions. This can prevent catastrophic failures and extend the life of the system.

- Model-based reasoning: It is useful to build detailed models of system behaviors to analyze, diagnose, or predict outcomes in different scenarios. These help to understand and reason about their environment, leading to more intelligent and adaptable algorithms.

Key Terminology (Figure 4.1)

- **Anomaly:** An anomaly is an occurrence where the performance or behavior of a system or component deviates from its expected or designed function. It can range from minor deviations to significant unexpected events and may not always lead to failure.

- **Failure:** A failure is a condition where a component or system no longer performs its required function. This may occur due to internal faults such as component breakdowns or external factors such as space debris impacts.

- **Fault:** A fault is a defect that can lead to system failure. Defects can be physical (such as cracks) or logical (such as bugs).

- **Root cause:** The root cause is the fundamental underlying issue that led to the fault and subsequent failure. Its identification is crucial for preventing future failures.

- **System:** A system refers to a set of interconnected components designed to achieve specific objectives, such as propulsion, communication, or navigation.

Figure 4.1 Terminology concept diagram.

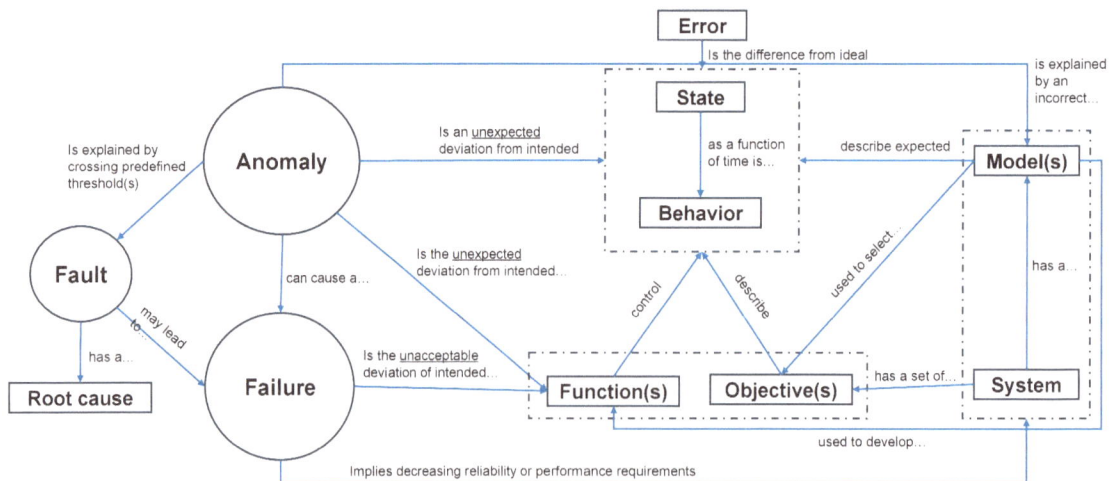

© SAE International

- **State:** The state refers to the current conditions or values of its various components and variables at a time. This can include positions, velocities, temperatures, and operational modes.

- **Behavior:** Behavior refers to how the state of the system changes. This includes the dynamic response of the spacecraft to inputs, control actions, and environmental conditions.

- **Function:** A function is the intended action or set of actions performed by a system or component, such as thrust generation in engines or data transmission in communication systems.

- **Error:** The error is the difference between the estimated state (based on sensor readings and models) and the desired state, as defined by the control objectives. This can be because of inaccuracies in the control system or unexpected external influences.

4.1.
FTA

Understanding how systems fail (and the complex ways in which failures happen) has led to the development of various approaches. One of these is called the deductive, top-down graphical analysis. It helps us see and understand failure paths and chains in complex systems. Consider a spacecraft battery system and its potential failures and causes. A deductive graphical analysis can give us a rational way to do this visually. It is like a map that shows us the different ways the system can break down.

Instead of focusing on the system, we adopt a "top-down approach" by analyzing its individual components and occurrences. This helps us not only find potential ways the system can fail but also trace the paths that lead to those failures.

This technique has a pretty interesting origin story. It was first developed at Bell Telephone Laboratories for the United States Air Force (USAF) back in 1962 who were interested in understanding the failure of complex telecommunications systems. This method became popular because it effectively investigates how different aspects of a system depend on each other and identifies potential vulnerabilities. Deductive graphical analysis is a versatile tool that finds its place in various fields, such as operations research and systems reliability. It is especially useful when we want to see the chain reactions that lead to failures in complex systems. It also works well alongside methods such as reliability block diagrams (RBDs) and fault trees. RBDs help us understand how different parts contribute to the overall success of a system, while fault trees focus on understanding failures.

A famous technique within deductive graphical analysis is creating fault trees. Consider fault trees as visual representations that depict the overall failure of a system. They are like roadmaps that highlight which parts can cause the complete system to fail. Constructing fault trees involves building them using logic gates that establish the relationships that could lead to failure. The fault tree diagram depicted in Figure 4.2 illustrates how a single event can lead to the failure of the entire system.

Figure 4.2 FTA of the (a) O-ring construction and (b) Challenger disaster.

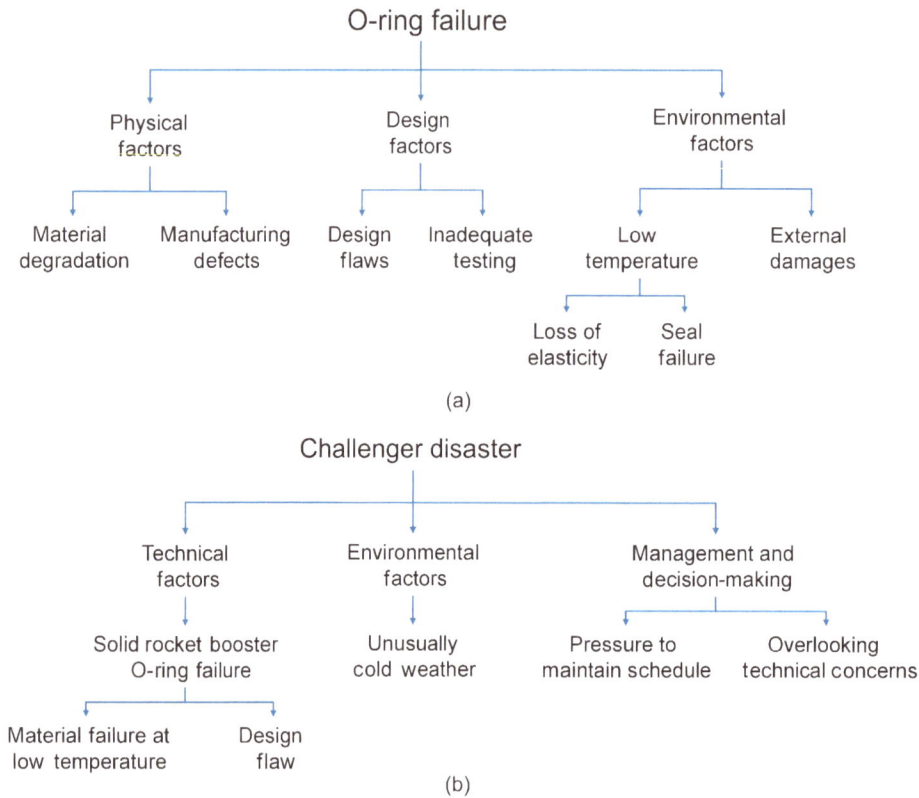

(a)

(b)

© SAE International

Adding an additional sensor to a system could also help investigate if it could result in an improved likelihood of achieving successful outcomes. The reason behind this counterintuitive observation lies in the subtleties of "redundancy design." Typically, we expect redundancy to enhance the robustness and reliability of a system. By introducing duplicate components or measures, the system can continue functioning even in the event of a single failure. Yet, in sensor redundancy, the key lies in the type of failure being considered. When integrating a redundant sensor, it often duplicates the measurements provided by the primary sensor. In scenarios where the primary sensor is functioning accurately, the redundant sensor

might generate virtually identical readings. However, the critical aspect arises when the primary sensor malfunctions or encounters an error. Where the primary sensor provides incorrect readings because of a malfunction, the redundant sensor is likely to replicate these inaccurate measurements.

This redundancy does not offer a beneficial effect on the overall system performance; rather, the correlated failure of both the primary and redundant sensors can lead to compounded errors and adversely affect the accuracy and reliability of the system's output. Effective redundancy design requires careful consideration of failure modes, independence of components, and fault detection mechanisms. Redundancy is most valuable when it involves diverse sensors that can independently validate each other's outputs. This way, the system can

intelligently choose the correct reading while identifying and discarding erroneous data (Figure 4.3).

Figure 4.3 Components for fault management.

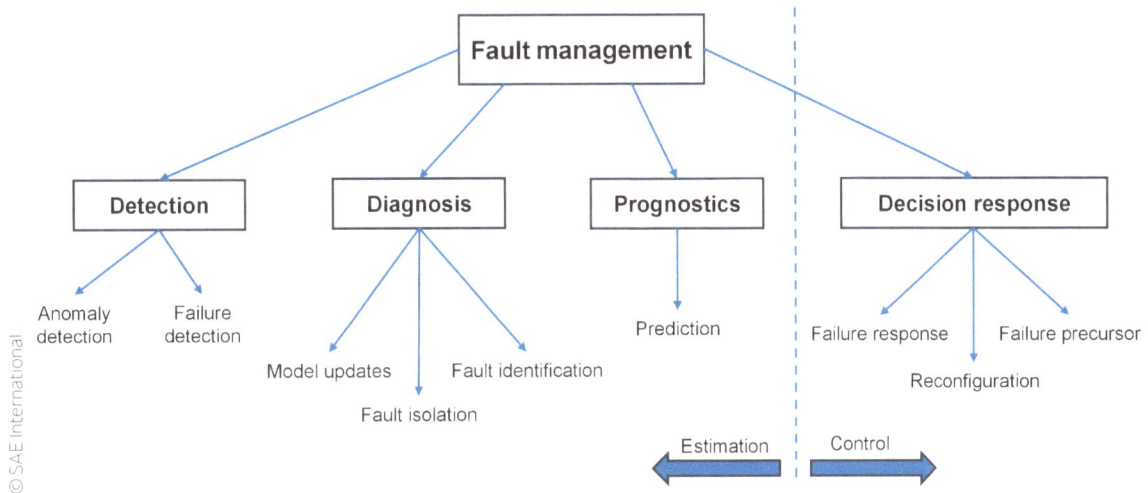

4.2.
Diagnostics Technology

Diagnosis involves identifying a particular issue. In the realm of health management, this shows a fault. The significance of diagnostics becomes apparent when considering real-world scenarios. For example, in a rover mission, the primary aim is to navigate and visit specific waypoints to fulfill various scientific goals. However, the mission's success hinges on the smooth operation of the rover's intricate systems. Despite meticulous planning and engineering, faults can still emerge because of the unpredictable nature of space environments and the complexity of the

machinery. Now, imagine a critical fault occurring during the mission. This fault could manifest as an unexpected change in the rover's behavior or functionality. Different factors might trigger it, spanning from HW malfunctions to SW glitches. Such faults, if left unaddressed, could lead to a catastrophic system failure, jeopardizing the entire mission. This is where diagnostics come into play. Diagnostics involve identifying the nature and root cause of the fault. In the rover mission, diagnostics would involve assessing which system or component has malfunctioned and what triggered the fault (Figure 4.4). Pinpointing the exact fault allows mission controllers to make informed decisions about the next steps. Here is where the complexity intensifies with the rover on a distant planet: no real-time communication with Earth. The vast distance introduces communication delays, making it impossible to diagnose and troubleshoot the fault remotely from Earth. In such scenarios, the rover must possess onboard diagnostic capabilities to detect anomalies and determine the reasons behind the fault. The rover can independently diagnose and respond, even without communication with Earth. Prognostics involves predicting the remaining useful life (RUL) of systems to enable timely interventions [4.1].

Figure 4.4 Simplified diagnostic loop.

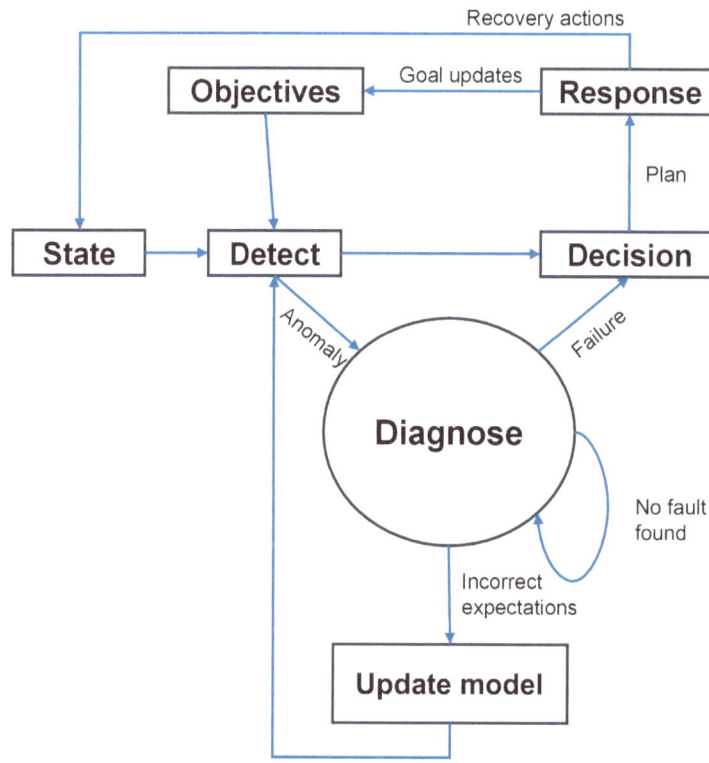

© SAE International

Diagnostics play a crucial role in guiding engineering decisions by providing valuable insights. This helps determine the components to fix or replace, helps effectively manage faults, aids in recovering from glitches, and even informs the reshuffling of functions if necessary. Diagnostics go beyond the present moment—it helps us predict the future. It shows wear and fault tolerance in operations. In a nutshell, diagnostics is like a strategic advisor, helping us make informed choices and ensuring the smooth functioning of systems.

Distinguishing between "nominal" and "faulty" behavior is essential in diagnostics (Figure 4.5). "Nominal" behavior refers to the expected functioning of a system based on established knowledge and reference standards.

This understanding stems from expert insights, known operational boundaries, physical models, ML techniques, and more. In the realm of model-based diagnostics, experts derive the reference for nominal behavior from a model that precisely outlines the anticipated behavior. Models can be static or dynamic representations, serving as valuable tools for simulating and understanding system operations under normal conditions.

Figure 4.5 Nominal vs. faulty behavior.

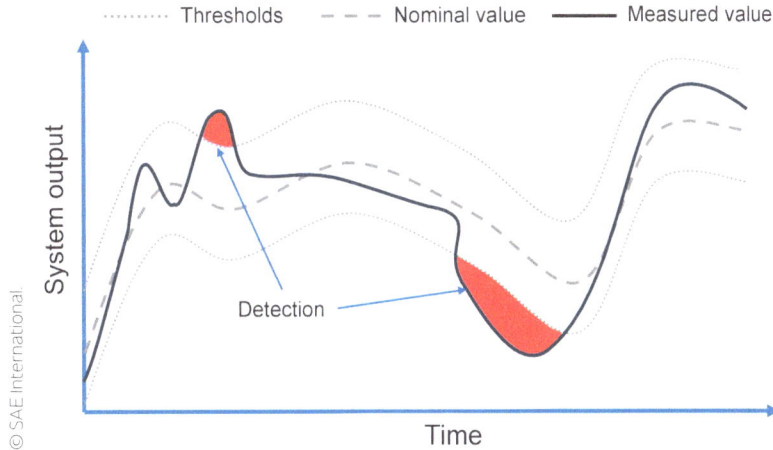

Distinguishing between "nominal" and "faulty" behavior is crucial to diagnostics. Now, *why do we opt for model-based diagnostics*? One interesting reason lies in the explanatory power that models bring to the table. These models enable us to engage in causal reasoning, understanding not just the outcome but the underlying reasons behind it. They also provide an explicit framework for representing faults, shedding light on potential deviations from the desired behavior. The power of models extends to the development of general model-based algorithms. These algorithms work with the models as inputs, acting as dynamic tools for diagnosing system issues. What is helpful is that the algorithms remain consistent regardless of the specific system under consideration. This approach has the advantage that we do not need to completely change the algorithm for new systems. This level of versatility streamlines the diagnostic process, making it more efficient and adaptable to various scenarios.

- **Fault detection:** Imagine a scenario where you are trying to find out whether a system is operating as it should be. It involves analyzing the system's behavior and comparing it to the anticipated "nominal" behavior we discussed earlier. If it detects any deviations or abnormalities, it raises a flag that something might be amiss.

- **Fault isolation:** If a system is not meeting the anticipated performance, it is crucial to pinpoint the precise root cause of this unforeseen behavior. Through the process of fault isolation, one can pinpoint the precise root cause of this unexpected behavior. It is like solving a puzzle—we are trying to figure out which specific component or aspect of the system handles the deviation from the norm.

- **Fault identification:** And finally, when we identify a fault, we often want to know more. *How severe is the fault? What is its magnitude?* This is where fault identification comes into play by quantifying fault extent for maintenance decisions.

Each capability has its own challenges, and model-based approaches provide a systematic way to address them, improving our ability to diagnose and resolve issues in complex systems.

- **Fault characterization:** Fault characterization is like understanding the unique personalities faults can have. Recognizing these variations equips us with the insights needed to tailor our diagnostic strategies effectively and address a wide range of fault scenarios.

- **Abrupt faults:** Picture a scenario where there is a rapid change in a parameter value. This change happens so swiftly that it is even faster than the system's sampling frequency.

These are what we call abrupt faults. They can arise because of sudden failures or disturbances and are often relatively easier to detect because of their sudden nature.

- **Incipient faults:** These involve changes in parameter values that occur more gradually, at a pace slower, than the system's sampling frequency. These can exhibit various patterns—linear, exponential, or even arbitrary degradation. When we talk about prognostics, which relates to predicting future system behavior, we are often referring to incipient faults. Detecting and predicting these types of faults can be challenging because of their subtle and gradual nature.

Interestingly, the dynamics of faults can differ from the dynamics of measurements or observations. For instance, an abrupt fault might lead to an incipient change in measurements (Figure 4.6). This distinction is essential because our diagnostic methods need to account for these variations in fault behavior.

Figure 4.6 Abrupt vs. incipient behavior.

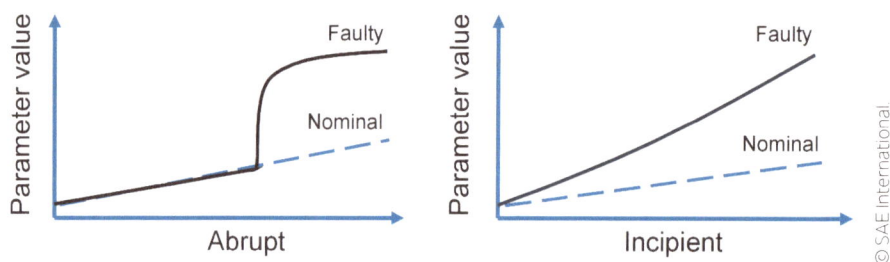

- **Persistent vs. intermittent faults:** When we say a fault is "persistent," it means that once it shows up, it sticks around—it is consistently affecting the system's behavior (Figure 4.7). On the flip side, an "intermittent" fault

is a bit more unpredictable. It manifests itself intermittently, causing disruptions at unpredictable intervals. Think of it like a flickering lightbulb—sometimes it works fine, other times it goes out.

Figure 4.7 Persistent vs. intermittent behavior.

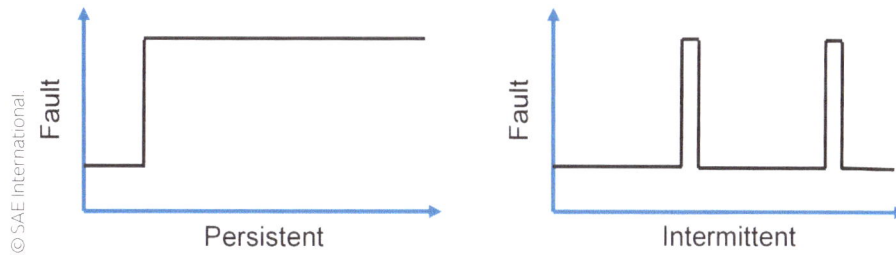

Discrete vs. parametric faults: Discrete faults involve unwanted changes in the system or model structure. Parametric faults relate to changes in the parameters of the system or model.

These distinctions—between persistent and intermittent faults, as well as discrete and parametric faults—help us paint a clearer picture of the complexity that underlies fault behavior. Differentiating between these fault types enables us to tailor our diagnostic strategies to address the unique challenges posed by each scenario. Whether it is a consistent glitch, a sporadic hiccup, a structural change, or a parameter shift, understanding these nuances equips us to diagnose and manage faults effectively.

4.2.1.
Continuous Diagnosis

Continuous system diagnosis combines fault detection, isolation, and identification into a comprehensive strategy. It focuses on singular, persistent parametric faults, which include both abrupt and incipient variations commonly seen in dynamic systems. The method hinges on understanding how changes in system parameters affect outputs, even without direct modifications to these parameters.

A well-structured model predicts how different faults might alter measurements. This model acts as a guide, showing the expected measurement shifts in various fault scenarios. As we navigate through these, the sequence of updated measurements provides clues about the fault's nature and progression. The key is aligning the deviations with those predicted by the model. Comparing actual deviations can identify the actual fault. This process, though complex in its need for precise model construction and temporal analysis, offers a thorough way to understand and navigate the intricate interactions in complex systems. It effectively links dynamic behaviors, parameter changes, and measurement deviations, providing a clear path in continuous system diagnosis.

The process of residual generation is also a critical step (Figure 4.8). It begins with the use of an observer, often based on methodologies like Kalman filters (KFs), unscented Kalman filters (UKFs), or particle filters (PFs). This observer, integrated within the nominal local submodel, creates a reference trajectory reflecting the system's expected behavior under ideal conditions. The core of this approach is calculating residuals, the difference between actual measurements, and the reference values from the observer.

By comparing these, we can gauge how far the system deviates from its norm. These residuals are more than just measurements; they offer insights into the system's behavior, highlighting anomalies that might otherwise remain hidden. The significance of residuals extends to fault detection. Under normal operation, residuals usually stay close to zero. However, significant deviations from this baseline show a departure from normal functioning, pointing to potential faults. It is important to note that fault detection is not immediate. There is an inherent delay between a fault's occurrence and its detection, caused by the system's dynamics. A thoughtful design and calibration can minimize the impact of this delay.

Figure 4.8 Model-based residual generation for fault detection.

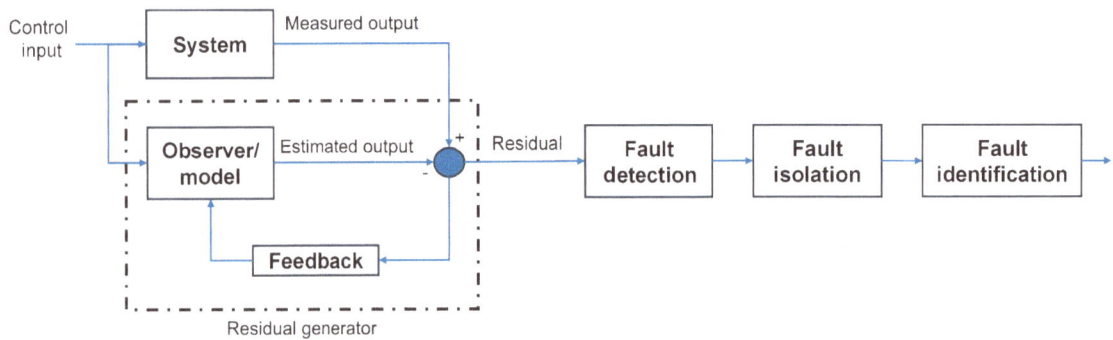

© SAE International

4.2.2.
Fault Isolation

In complex systems, accurately identifying the specific nature of faults is crucial. This process relies on comparing observed discrepancies in measurements to expected fault signatures, which can uniquely determine each fault based on the specific deviations it causes. This concept becomes even more critical in systems with multiple faults and measurements. In such a system, with a range of distinct faults ($f_1, f_2, ..., f_m$) and various measurements ($m_1, m_2, ..., m_n$), we can determine the diagnosability of the set F by isolating each fault using the measurements. Each fault in the system must have a distinct fault signature. This methodical approach to fault isolation is essential for pinpoint accuracy in fault identification. This ensures that we recognize and categorize even the most subtle differences in faults correctly, highlighting the importance of precision and thoroughness in system diagnostics.

4.2.3.
Fault Identification

The challenge of parameter estimation in fault identification involves determining the exact value of a new fault parameter after observing abnormal system behavior. This complex task employs various algorithms, each designed to solve the parameter estimation challenge effectively. One common strategy is to define an estimation window. This is chosen carefully to include data from just before the fault detection time until now. Within this window, the mechanism for observation is activated. This observer works alongside the state vector, which now includes the fault parameter, enabling a joint estimation of both the state and the parameter. This method provides a comprehensive view of the system's current state and the nature of the fault.

Another innovative approach involves developing a submodel that represents the unknown fault parameter as a function of known, measurable variables. This method delves into the relationship between known factors to deduce the elusive fault parameter. It offers a different perspective, focusing on the connections between established variables to uncover the unknown.

4.2.4.
Fault Protection

Robust design features can ensure operational integrity and system protection. The scope often includes concepts such as fault avoidance, fault tolerance, and fault masking. Engineers implement these strategies in various combinations, depending on the mission requirements, spacecraft design, and the specific risks associated with the mission. The goal is to ensure mission success despite these inherent risks and challenges and adapt to unexpected conditions and failures, enhancing the overall reliability and safety of space missions.

4.2.5.
Fault Avoidance

It is a strategy that aims to prevent errors before they occur. This is crucial in space missions where repairing a fault postlaunch is often impossible or highly impractical. Fault avoidance encompasses several key practices and design philosophies:

- Robust design features: This involves creating spacecraft components and systems that are inherently reliable and less prone to failure. Robust design includes:

  - Simplicity in design: Implementing designs with fewer moving parts or complexities reduces the likelihood of failure. For example, using fixed solar arrays or high-gain antennas (HGAs) minimizes mechanical failures as fewer components can malfunction.

  - Conservative design practices: This includes designing systems with significant performance margins. Engineers often design spacecraft components to withstand conditions far more extreme than their expected encounters. This ensures that, even under unexpected circumstances, the system remains functional.

- Onboard autonomy: Spacecraft are often millions of miles away from Earth, making real-time intervention by ground control impractical. Therefore, they are equipped with autonomous systems capable of making critical decisions independently. For instance:

  - Attitude constraint checking: This system autonomously adjusts the spacecraft's orientation to maintain optimal positioning, such as ensuring that the sun adequately exposes the solar panels and correctly aligning the antennas for communication.

- Postlaunch operations processes and procedures: After launch, the focus shifts to operational strategies that prevent faults during the mission. These include:

  - Use of flight rules: These are predefined guidelines that dictate how the spacecraft should operate under various conditions. Flight rules are based on extensive simulations and historical data, ensuring that the spacecraft responds optimally in different scenarios. Process history-based methods are essential for fault detection and diagnosis, providing insights into recurring faults across similar systems [4.2].

- Spacecraft simulation testbeds: Engineers frequently test commands on ground-based simulators before executing them on the actual spacecraft. This shows how the spacecraft will react to certain commands and identify any potential issues before they occur on the actual mission.

4.2.5.1.
Fault Tolerance

Fault tolerance allows systems to continue operating in the event of a system failure. Space missions require this capability as external help is limited or nonexistent. The key strategies in fault tolerance include several well-established practices:

- Graceful degradation: This strategy allows a spacecraft to maintain operational capability at a reduced level after a component failure. Instead of a complete system shutdown, only the affected component ceases to function, while other systems continue operating. For example, if one of several onboard sensors fails, the spacecraft can still complete its mission using the remaining sensors, albeit with reduced data collection capabilities.

- Redundancy:
  - Physical redundancy: Involves having duplicate critical components within the spacecraft. If one component fails, its backup can immediately take over, ensuring continuous operation. Many spacecraft designs incorporate physical redundancy, such as including multiple computers or redundant power systems.
  - Functional redundancy: Different systems can perform the same function. This type of

redundancy ensures that, if one system fails, another distinct system can take over its function. An example could have both chemical and ion propulsion systems on a spacecraft.

- Fault containment regions (FCRs): These prevent a fault in one part of the spacecraft from affecting the entire system. FCRs are like firewalls within the spacecraft's architecture, isolating problems to prevent cascading failures. For instance, if one module of a spacecraft experiences a fault, the fault can be isolated to prevent it from spreading to other critical systems.

- Fault detection, isolation, and recovery (FDIR): It provides a structured approach to detect and isolate faults, ensuring system reliability and safety [4.3].

  - Fault detection: The system continuously monitors various spacecraft parameters to detect anomalies that could show a fault.

  - Fault isolation: Once the system detects a fault, it determines the exact location and nature of the fault. This step is crucial for implementing effective recovery procedures.

  - Fault recovery: After isolating the fault, the system starts predefined procedures to recover from the fault. This could involve switching to a redundant system, reconfiguring the system to bypass the fault, or even changing the spacecraft's operational mode to mitigate the impact of the fault.

4.2.5.2.
Fault Masking

Fault masking is a critical strategy designed to conceal the effects of faults, ensuring that the overall system continues to function correctly despite errors or malfunctions. This approach is important in space missions, where even minor faults can have significant consequences. Fault masking involves several key techniques:

- Error detection and correction (EDAC): EDAC systems are crucial for managing data integrity, especially in communication systems. They detect and correct errors in transmitted data, which is essential in the space environment where radiation can cause bit flips and other data corruptions. For example, the communication system of a spacecraft might use EDAC to ensure that the data received on Earth are accurate, despite the harsh conditions of space.

- Field-programmable gate array (FPGA) triple modular redundancy (TMR): In FPGA-based systems, TMR involves triplicating logic circuits and then voting on their outputs. This method is effective in mitigating single-event upsets (SEUs), which are common in space because of cosmic rays and other radiation sources. If any circuit gets affected, the other two can outvote it, masking the fault.

- Active redundancy: This involves having multiple identical systems running in parallel. Unlike its passive counterpart where backup systems are activated only when the primary system fails, these systems constantly compare their outputs, and if one diverges (indicating a fault), they ignore or take it offline, while the remaining systems continue to operate. Engineers often use this approach in critical spacecraft systems such as navigation and control, where continuous operation is essential.

- Watchdog timers: These are used to reset a system automatically if it becomes unresponsive or behaves erratically. A watchdog timer continually monitors the system's operation. If the system does not perform a specific action within a predetermined time frame (showing a potential fault), the timer will reset the system, effectively masking the fault and allowing the system to recover.

- Built-in self-test (BIST): Many spacecraft systems include BIST capabilities, which allow them to perform routine checks on their operation and correct any detected faults. This can involve running diagnostic algorithms that can detect and compensate for certain types of faults, effectively masking them from the rest of the system.

4.3.
Model-based Reasoning

In the past, experts have suggested many approaches to identify model parameters using measured data, such as regression-based approaches [4.4]. However, the most popular technique used stems from Bayesian statistics, applied in various ways to identify parameters. The well-known algorithms, such as the PF, KF family, and Bayesian method, all rely on integrating a physical damage model with system data to forecast future damage or degradation behaviors. The mathematical model characterizes the observed system, enabling failure and degradation analysis. By simulating the model alongside actual processes, we can detect deviations from normal operations. Constructing such a model causes an in-depth understanding of the underlying physics, leading to complex formulations as system complexity increases. Although these models entail inherent assumptions and approximations, model validation is essential before application. Researchers derive parameters for the model's behavior from laboratory tests or real-time data estimation. A notable advantage is their minimal data dependency and independence from data during specific failures, which is often needed in supervised training methods. In his 1987 General Diagnostic Engine, de Kleer used a diagnostic reasoning approach to lay the foundation of the concept by reconciling observed and theoretical values from a physical model. Residuals between actual sensor values and model-predicted values quantity errors, with the KF as a prime example, estimating system states amid noisy data. Despite assuming Gaussian noise and imposing linear equation restrictions, researchers have extended this filter through alternatives such as PFs, Bayesian methods, and extended Kalman filters (EKFs). These alternatives are used to address nonlinear relationships and non-Gaussian noise.

There are several critical challenges:

- The definition of a degradation signal, serving as a component's health indicator. It must capture varying stages of degradation and be detectable by engineering sensors. Constructing robust health indicators requires a thorough understanding of component physics, encompassing degradation mechanisms, potential faults, and mutable parameters. Selected signals that accurately mirror degradation processes and are sensor-friendly form the foundation for a comprehensive health indicator, synthesized through methods.

- Establishing a degradation model, which reflects the degradation processes. Although parameter estimation for this model may be difficult due to complex degradation, empirical models based on observed signals are often used.

- Understanding uncertainties, given the core role of predicting future degradation evolution and RUL. These uncertainties stem from unit variances, stochastic operational loads, measurement errors, and model limitations. While full elimination of uncertainties is unattainable, their transparent articulation, quantification, and incorporation within forecasts enhance the reliability and precision of prognostic outcomes.

In various engineering contexts, degradation phenomena often conform to linear processes featuring a linear drift. This implies that the mathematical depiction of degradation involves the accumulation of deterioration at a consistent pace. As a result, a straightforward model becomes instrumental in illustrating the approach to fault prediction. To compute the RUL, it is essential to establish a fault threshold denoting the highest tolerable level of degradation. Once the degradation surpasses this fault threshold, we categorize the component as faulty. The RUL value is not deterministic, fluctuating with variations, and hence the fundamental aim here is to compute the cumulative distribution function (CDF) of RUL. We can approximate the degradation process as an incremental process with a specific rate. Due to factors such as material, assembly, and degradation stress, the change in each degradation increment can be modeled as a normal distribution, following the central limit theorem in statistics. Thus, viewing the degradation process from this perspective, it takes on the characteristics of a random-effect Wiener process. In this context, a Bayesian framework becomes a powerful tool for handling uncertainties, particularly in prediction. Notably, methods such as Kalman filtering and particle filtering play significant roles.

4.3.1.
Kalman Filtering

State estimation is a fundamental problem in various fields, from robotics to finance and aerospace. The primary goal is to estimate the internal states of a system, given noisy measurements and a model of the system's dynamics. Over the years, researchers have developed several filtering techniques to address this problem, with the KF, EKF, and PF emerging as the most prominent ones.

The KF, introduced by Rudolf E. Kalman in the early 1960s, is a recursive algorithm designed for linear systems with Gaussian noise. It provides an optimal solution under these conditions and operates in two key steps: prediction and update. The prediction step projects the current state estimate forward in time, while the update step refines this prediction using a new measurement. Many applications have widely adopted the KF because of its simplicity, optimality for linear Gaussian systems, and low computational cost.

Assuming a reliable physical model is available, the next major challenge involves carrying out model parameter identification using an estimation algorithm based on measured data.

Suitable for linear models with Gaussian distribution, Kalman filtering aims to establish a probability density function (PDF) for the state based on measurement knowledge encompassing state and model parameter identification. By utilizing state equations, past estimates, and a sequence of measurements affected by noise from various sensors, the algorithm generates an optimized state estimate for a system. The algorithm iteratively employs Bayesian inference to maximize probabilities of measurements and states.

The KF algorithm operates through two key steps:

- Prediction: The model that represents the system's underlying dynamics projects the state vector forward, along with its associated uncertainty.
- Measurement: The system adjusts the projected state based on actual sensor measurements, yielding an optimal estimate.

The weights assigned to measurements and predictions consider their respective uncertainties. However, the EKF addresses the assumptions of the KF by linearizing both the sensor measurement and system dynamics models. This entails calculating the Jacobian matrices containing partial derivatives for these models (Figure 4.9).

Figure 4.9 Block diagram of KF.

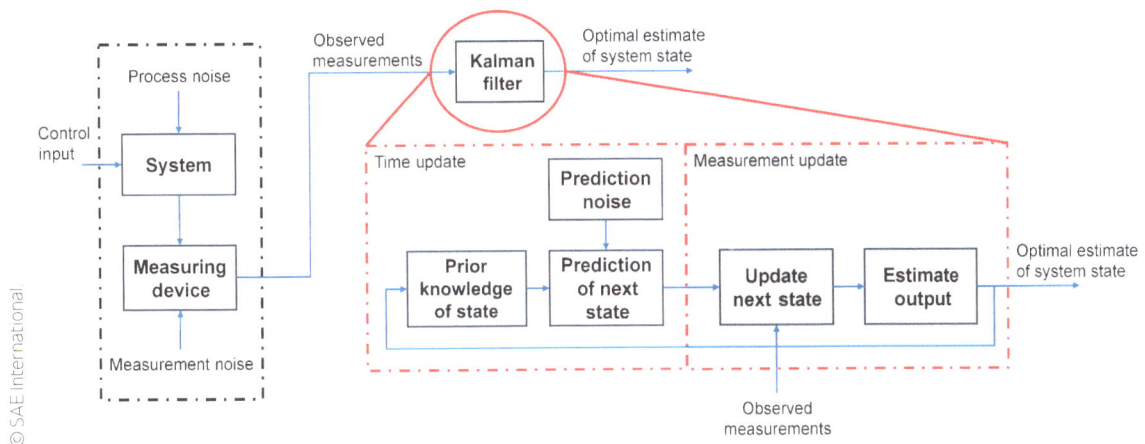

© SAE International

4.3.2.
Extended and Unscented KF

While the KF is optimal for linear systems, many real-world systems exhibit nonlinear behavior. The EKF extends the KF that addresses this nonlinearity. Instead of directly dealing with nonlinear systems, the EKF linearizes the system dynamics and measurement functions about the current state estimate using Jacobians. This allows the EKF to apply the standard KF equations to the linearized system. However, EKF's performance can degrade if the system's nonlinearity is strong, leading to potential estimation errors. In the EKF, the state transition and observation models may be nonlinear functions of the state, as long as they are differentiable.

The EKF algorithm's simplicity and robustness have made it a widely used technique for estimating nonlinear system states. However, the EKF has two primary limitations that can complicate practical implementation in certain cases. First, highly nonlinear state transition models can lead to instability and suboptimal performance. Second, the EKF's reliance on a first-order truncated Taylor series for linearization imposes limitations on the accuracy of the propagated covariance and mean estimates beyond the first order.

Unlike its linear counterpart, the EKF is not an optimal estimator unless both the measurement and state transition models are linear. In such cases, it becomes equivalent to the regular KF. If the initial state estimate is incorrect or if the process model is flawed, the filter's linearization can lead to divergence. Without introducing "stabilizing noise," there is a risk of the estimated covariance matrix underestimating the true covariance matrix and leading to statistical inconsistency. Despite these challenges, many navigation systems and GPS widely use the EKF because of its ability to yield reasonable performance, making it a de facto standard.

Researchers created alternative versions of the KF, like the UKF, to handle the challenge of high nonlinearity in process and measurement models, in response to the limitations of the EKF. The UKF achieves this by generating derivative-free estimations with Gaussian distributions, making it more robust and suitable for nonlinear systems. The UKF employs the unscented transform (UT), a central component that computes statistics for a nonlinear transformed random variable. It characterizes the state vector and its associated uncertainty using "sigma points," which are used to approximate distribution moments. The UT approximates nonlinear transformations using a probability distribution defined by a finite set of statistics. This approach is useful for projecting covariance and mean estimates into nonlinear extensions of the KF.

The UKF demands more computational resources than the EKF, but it compensates for this with superior performance. This enhanced performance stems from its more accurate moment approximations, thanks to the UT process. The UKF's accuracy is comparable to that of the second-order EKF, which requires computing both Jacobian and Hessian matrices. Despite KF techniques being well-established, they continue to inspire innovative applications in diagnostics and prognostics.

4.3.3.
PF

The PF is a distinctive approach that stands out as a sequential Monte Carlo (SMC) method. It is designed to create a recursive Bayesian filter through Monte Carlo simulations.

Unlike methods that rely on Kalman gain or adjoint sensitivity techniques, such as the KF and its variations, the PF takes a different route. In this method, a collection of particles is used to represent the posterior distribution of a stochastic process. It could be cracks propagating due to damage. Starting with initial observations, the PF uses this particle-based representation to estimate the state that is most likely to occur next. One of the key strengths of the PF is its ability to handle nonlinear state-space models. It offers considerable flexibility when dealing with noise distributions and initial states, making it adaptable to a variety of complex scenarios.

An overview of the PF process steps include:

- Prediction: The corresponding CDF multiplies the PDF of the previous state-space posterior distribution. Then, we integrate this over the states from $(n - 1)$ to obtain the prior particle distribution for the current step.

- Update: The probability from measured data (represented as weights assigned to each particle) is used to update the state vector.

- Resampling: Particles in the updated distribution are resampled based on their weights. This involves replicating particles with high weights and removing particles with low weights.

These steps contribute to the recursive estimation process of the PF, allowing it to approximate complex posterior distributions and track the evolving state of the system. PF algorithms are highly adaptable and effective for handling nonlinear systems, especially those with non-Gaussian distributions. One such application involved using a PF-based algorithm that integrated physical models for degradation and crack growth to predict the RUL of systems.

In another instance, the researchers developed an online PF-based framework specifically tailored for fault detection, identification, and failure prognostics in nonlinear and non-Gaussian systems. However, implementing PF algorithms in real-time applications presents certain challenges. A notable issue is the potential mismatch between the high-rate influx of sensor data and the update rate of the filter. To address this, some solutions, such as real-time particle filters (RTPFs), have proposed allocating samples to new observations during each filter update. There have been efforts to reduce the computational demands of PF-based failure prognostics, making them more suitable for real-time applications. Despite these advancements, implementing PFs comes with its own set of limitations. Key among these is the accuracy of the initial distribution and the accumulation of sampling errors over successive iterations. Another issue is particle depletion, which occurs when particles become unevenly distributed among states. Introducing random samples during the prediction phase can partially mitigate this, preventing particle duplication and elimination. However, this solution increases computational demands and can affect the probabilistic properties and parameter variance in the system.

4.4.
Summary

The importance of good engineering practices is critical in creating safer, more sustainable technologies through smart decision-making and innovative solutions. It emphasizes data analysis and pattern recognition as key to informed decision-making in SHM, blending systems engineering, reliability engineering, and data analytics. This method turns raw data into useful information, allowing for the prediction and control of complex systems' well-being. Key methods discussed include FTA for understanding system vulnerabilities, diagnostics for identifying current system states, prognostics for future health predictions, and model-based reasoning for outcome analysis. These concepts are crucial for improving system reliability and maintenance, showcasing the integration of data-driven insights with engineering practices to enhance technological advancements and operational efficiency.

References

4.1. Lee, J., Wu, F., Zhao, W., Ghaffari, M. et al., "Prognostics and Health Management Design for Rotary Machinery Systems—Reviews, Methodology, and Applications," *Mechanical Systems and Signal Processing* 42, no. 1-2 (2014): 314-334.

4.2. Venkatasubramanian, V., Rengaswamy, R., Yin, K., and Kavuri, S.N., "A Review of Process Fault Detection and Diagnosis: Part III: Process History-Based Methods," *Computers & Chemical Engineering* 27, no. 3 (2003): 327-346.

4.3. Isermann, R., *Fault-Diagnosis Systems: An Introduction from Fault Detection to Fault Tolerance* (Berlin, Heidelberg: Springer, 2006).

4.4. Saxena, A. and Goebel, K., "PHM: A Comprehensive Review of the State of the Art," in *Proceedings of the 1st International Conference on Prognostics and Health Management (PHM2008)*, Denver, 2008.

We understand space missions involve significant investments of time, resources, and human capital. Therefore, mission success requires guaranteeing the reliable functioning of spacecraft systems, whether it involves deploying satellites, conducting scientific experiments, or exploring space with humans. In the previous chapters, we have discussed some unfortunate events and challenges associated with the harsh operating environment and inherent system complexities. These introduce uncertainties in known models. For example, understanding how damage behaves relies heavily on accurately identifying the model parameters. We also need to figure them out to predict future behaviors. Because of the uncertainties and noise in data, most algorithms incorporate model parameters as a range of possibilities, not just fixed values.

This chapter discusses health management systems in spacecraft. It also touches on the modern view of integrating certain capabilities for improving results.

Health management technology in spacecraft involves monitoring system anomalies through signals, models, and algorithms to predict failures and determine maintenance strategies. This proactive approach aims to prevent faults before they occur, evaluating system health degradation and predicting the remaining life.

Typically, spacecraft would employ:

1. Telemetry data: In the early days of space exploration, spacecraft health monitoring primarily relied on telemetry systems. These systems would transmit data from the spacecraft's onboard sensors to ground stations on Earth. Engineers and scientists would then analyze these data to assess the spacecraft's health and performance.

2. Predefined thresholds: The monitoring systems were often based on predefined thresholds. If a particular parameter (such as temperature, voltage, or pressure) exceeded or fell below its designated threshold, the monitoring systems would generate an alert. This method was straightforward but could not capture complex or unforeseen anomalies.

3. Ground-based analysis: This involves conducting most of the data analysis on the ground. After the spacecraft sends the raw data, scientists on Earth would process and analyze them. This approach required a constant communication link and could introduce delays in detecting and addressing issues.

4. Redundancy: Given the high stakes of space missions and the inability to conduct repairs in space, redundancy was a key feature. Critical systems often had backups, so if one system failed, its backup could take over. This approach increased the spacecraft's weight and complexity but was essential for ensuring mission success.

5.1.1.
Telemetry Data

Telemetry, at its core, is an automated communication process. It involves collecting data from remote instruments, processing these data, and then transmitting it to a receiving endpoint for monitoring. Within the context of a spacecraft,

this translates to a myriad of sensors strategically placed throughout the craft. These sensors continuously monitor a plethora of parameters, ranging from temperatures and pressures to voltages and the status of various components. Once the spacecraft collects these data, its onboard computer systems process them and prepare them for transmission back to Earth. Onboard antennas and communication systems facilitate the transmission of these data, either in real time or by storing them onboard for transmission at scheduled intervals. Back on our planet, ground stations equipped with large antennas stand ready to receive these data. These stations, strategically scattered across the globe, ensure continuous communication with the spacecraft, regardless of its position in space.

Once the data reach Earth, they undergo further processing and analysis. Teams of dedicated engineers and scientists use advanced algorithms and SW tools to search for any anomalies or potential issues. This continuous stream of data provides an opportunity for predictive maintenance by foreseeing potential failures and taking preventive measures, prolonging the operational life of the spacecraft. This not only ensures the success of the mission but also proves to be cost-effective, reducing the risks associated with potential mission failures or extended recovery operations. Archived information offers insights into long-term trends, wear, and tear of components, and even providing valuable feedback for the design and operation of future spacecraft.

However, like all systems, telemetry-based health monitoring is not without its challenges. The sheer volume of data generated can be overwhelming, necessitating sophisticated data processing, analysis, and storage solutions. For missions venturing deep into space, significant communication delays can arise, potentially hindering real-time decision-making. The system's reliance on a network of ground stations introduces another layer of complexity, as a range of conditions can affect these stations, including adverse weather. Some examples include:

- Apollo missions.
  - Transmission of data related to spacecraft's position, velocity, and onboard system status.
  - Monitoring of astronaut's health and environmental conditions.
- Mars rovers (e.g., Curiosity, Perseverance).
  - Transmission of data from various scientific instruments and cameras.

- Monitoring of the rover's health, including power levels, temperature, and mobility system status.
- Satellites (e.g., HST, RADARSAT) (Table 5.1).
  - Transmission of scientific data from onboard instruments.
  - Monitoring of the telescope's health, including power levels, thermal conditions, and pointing status.
  - Transmission of synthetic aperture radar (SAR) imagery.
  - Monitoring of satellite health, including power levels, thermal conditions, and system status.

Table 5.1 Telemetry details.

Aspect	Apollo missions	Mars rovers	HST	RADARSAT
Data Volume	Apollo 11 transmitted approximately 16.2 kB/s.	Curiosity rover transmits up to 250 Mb of data per day.	Transmits approximately 120 GB of scientific data per week.	RADARSAT-1 transmitted approximately 45 GB of data per day.
Distance	Up to 238,855 mi (384,400 km) from Earth.	Up to 249 million mi (401 million km) from Earth.	Approximately 340 mi (547 km) above Earth.	LEO, approximately 600 km above Earth.
Mission duration	Apollo missions lasted from a few days to just over a week.	Spirit lasted 6 years, Opportunity lasted nearly 15 years, and Curiosity is ongoing.	Launched in 1990, still operational as of the last update in September 2021.	RADARSAT-1 operated for 17 years, RADARSAT-2 launched in 2007 and is still active.
Telemetry content	Included spacecraft health, astronaut biometrics, and mission-specific data.	Includes rover health, scientific data, images, and environmental conditions.	Includes telescope health, scientific data, and images.	Includes satellite health, radar imagery, and environmental data.
Challenges	Limited bandwidth and data loss during transmission.	Limited power for data transmission, must prioritize data to send.	Challenges with data transmission due to its distance and orbital position.	Requires precise antenna alignment for data transmission.
Historical significance	Its first mission to send humans to the Moon was significant for space exploration.	First rovers to explore Mars' surface, providing scientific data.	Has provided some of the most detailed images of distant galaxies and phenomena.	First Canadian satellite to use SAR technology.

5.1.2.
Predefined Thresholds

Historical data, engineering judgment, and sometimes simulations and testing are used to set these thresholds. These thresholds are used to monitor the real-time data from the spacecraft. If any of the parameters cross their respective thresholds, it triggers an alert or an automatic response to mitigate potential risks. The technical implementation involves continuous monitoring of critical parameters, such as temperature, pressure, voltage, and current. We set the thresholds by considering the safe operating limits derived from the components' specifications and testing data. When a parameter crosses its threshold, the system generates alerts for ground control or triggers automatic responses to address the anomaly. For instance, ground control quickly detects and addresses any deviations from the normal operating range of the Mars rovers by monitoring battery voltage and current. Similarly, temperature monitoring in satellites like the HST plays a crucial role in preventing damage to sensitive instruments.

The application of predefined thresholds in spacecraft SHM holds a rich historical significance, dating back to early missions such as the Apollo program, where NASA relied on it for real-time monitoring and ensuring astronaut safety. This method enables the spacecraft to respond autonomously to anomalies by comparing sensor data against set thresholds, triggering automatic responses to prevent damage or system failure. Despite its simplicity and proven reliability over decades of space exploration, the approach is heavily reliant on accurately setting the thresholds, necessitating an extensive knowledge of spacecraft systems and the harsh space environment. However, one limitation of predefined thresholds is their lack of predictive capability; they excel at anomaly detection but cannot foresee potential issues. In contemporary spacecraft health management, engineers often integrate these thresholds with advanced methods like model-based approaches and ML algorithms to enhance predictive capabilities and support system resilience. This integration is crucial for long-duration and deep-space missions, where early anomaly detection and mitigation are paramount for mission success. The adaptability of predefined thresholds to various spacecraft and missions, coupled with their significant contribution to the safety and success of space exploration, under-scores their role in health management.

To summarize:

- Pros:
  - Simplicity: The approach is straightforward to implement.
  - Real-time monitoring: It allows for real-time health monitoring and quick detection of anomalies.
  - Automatic response: Can be programmed to initiate automatic corrective actions, which are crucial in space missions where time is of the essence.
- Cons:
  - Sensitivity: Incorrectly set thresholds can lead to frequent false alarms or missed detections.
  - Lack of predictive capability: This method is reactive and does not predict future system states or failures.
  - Not adaptive: The thresholds do not adjust for spacecraft aging or environmental changes.

5.1.3.
Ground-based Analysis

Ground-based analysis refers to the monitoring and evaluation of health data from facilities on Earth. This approach relies on a continuous stream of telemetry data transmitted from the spacecraft to ground stations. The data can range from temperature and voltage parameters to system status logs, all of which are produced during flight. The process starts at telemetry receiving stations, where they capture and relay vast streams of data transmitted from the spacecraft to data processing centers. At these centers, the staff can decode, calibrate, and format the raw telemetry to transform it into actionable information. By using diagnostic tools, we investigate further, gaining a granular understanding of the system's health and pinpointing potential issues. This analysis presents the results to mission control teams, comprising experts who interpret the results, make informed decisions, and start necessary actions to address any identified issues. Through this layered approach, a ground-based analysis facilitates timely interventions and contributes to the success and safety of space missions.

Components of ground-based analysis:

- Telemetry receiving stations: These stations are equipped with large antennas and sophisticated equipment to receive signals from spacecraft. They play a pivotal role in ensuring a constant flow of data for analysis.

- Data processing centers: Once the telemetry data are received, the data processing centers forward them for decoding, calibration, and transformation into a format suitable for analysis.

- HMSs: These systems analyze the processed data, comparing them against predefined thresholds and historical patterns to identify any anomalies or deviations from expected behavior.

- Diagnostic tools: In the event of anomalies, experts employ diagnostic tools to pinpoint the source of the issue and assess its severity.

- Mission control teams: These teams comprise experts who interpret the analysis results, decide, and start actions to address any identified issues.

5.1.4.
Redundancy

Redundancy includes extra components or systems to take over if something fails. This approach stems from the understanding that space missions are high-stakes endeavors, where failure carries immense costs, not only in terms of financial investment but also in terms of scientific opportunity and human safety. There are two primary types of redundancy used in spacecraft health management:

- HW redundancy: This involves the inclusion of duplicate HW components. If one component fails, the other component can activate and ensure continuous operation. Critical systems such as power supply, communication, and propulsion commonly incorporate HW redundancy.

- SW redundancy: It involves implementing additional SW algorithms and routines that can be used if the primary SW fails or behaves unexpectedly.

Implementing redundancy in spacecraft is a complex yet crucial aspect of space mission design, involving careful planning and meticulous design. The addition of redundant components, while enhancing reliability, brings challenges such as increased weight and power consumption, both of which are critical considerations in space missions.

Adding extra components is necessary, as is preventing redundant components from failing like the primary ones. To achieve this, thorough testing and validation are essential. Automated decision-making processes often handle the switch to redundant systems, relying on predefined thresholds and algorithms to determine when a switch is necessary. This includes sophisticated monitoring and decision-making algorithms that enable autonomous switching to a backup system in the event of a failure.

Redundancy in spacecraft is not limited to HW. SW redundancy is equally important. It ensures that backup algorithms and routines can take over in case of an SW failure. This involves creating multiple layers of SW systems that can independently operate the spacecraft if the primary SW system fails.

However, managing redundancy in spacecraft also involves addressing the risk of common-mode failures, where both the primary and redundant systems fail because of a shared vulnerability. This requires careful design and rigorous testing to mitigate such risks. Another critical aspect is efficient resource allocation, as managing additional components and power for redundant systems is necessary. The evolution of spacecraft technology has led to more

sophisticated redundancy management systems. These systems are capable of nuanced decision-making, handling the complexities of modern space missions with greater efficiency. Instances of successful redundancy in preventing mission failure provide valuable lessons for future missions, continually improving the reliability of assets.

5.2.
Architecture Design

Architectures can be tailored by their complexity, adopting a centralized, distributed, or hybrid approach known as hierarchical and distributed. In a centralized architecture, a singular management controller oversees all health-related data, making it ideal for simpler spacecraft. Conversely, the distributed architecture decentralizes tasks such as monitoring, fault detection, and isolation, simplifying system integration and testing but at the cost of reduced reliability in fault diagnosis because of the lack of data fusion across subsystems.

To leverage the strengths of both centralized and distributed systems while mitigating their drawbacks, we can realize a hierarchical and distributed architecture for more complex spacecraft systems. These systems typically comprise multiple subsystems and components with varied functions and operational modes. This architecture stratifies the health management system into three levels: system, subsystem, and component, each responsible for specific tasks ranging from health data management to fault diagnosis and resolution based on a comprehensive knowledge database.

- At the system level, the architecture focuses on collecting, storing, and processing health data, interacting with ground systems based on the severity of health status, and coordinating with subsystems for fault management and reconfiguration. This level also integrates health data from various subsystems to resolve inconsistencies, accurately identify faults, and ensure reliable subsystem health assessments.

- The subsystem level is tasked with gathering and processing health data from its components, executing fault resolution processes at both the subsystem and component levels based on diagnostic outcomes and predefined knowledge models, and sharing health and fault resolution data with the system layer.

- The component layer addresses fault diagnosis and management at the most granular level,

using self-test or sensor data to execute fault management strategies informed by the knowledge model and communicating health and fault resolution information to the subsystem layer.

This hierarchical and distributed approach combines centralized oversight with distributed processing power, enabling a more nuanced and effective management of spacecraft health. It ensures that each layer—system, subsystem, and component—plays a pivotal role in maintaining the spacecraft's operational integrity, enhancing the overall reliability and efficiency of the spacecraft's autonomous health management system (Figure 5.1 and Table 5.2).

Figure 5.1 Health management architecture.

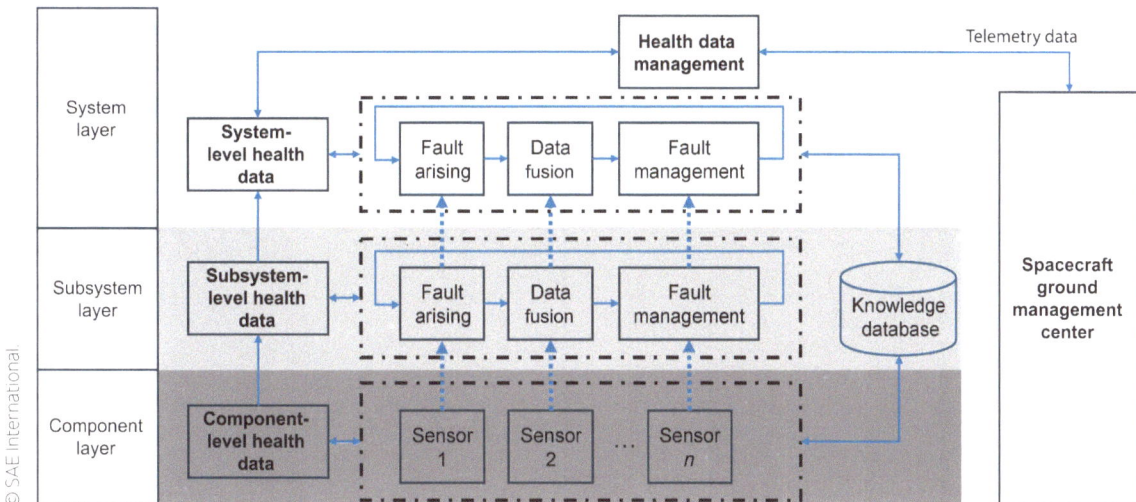

Table 5.2 Health management strategies in spacecraft systems.

Strategy	Strategy description	Implementation tools/ techniques	Expected outcome
System monitoring	Monitors the status of the intelligent control unit of each subsystem. In case of abnormalities, the system management unit starts recovery procedures for each intelligent control unit, such as reconfiguration, resetting, or switching between main/backup systems.	Sensors, anomaly detection algorithms, system management SW.	Enhanced reliability and uptime through proactive identification and resolution of subsystem issues.
Payload mission safety	Checks the validity of user mission commands, including task conflicts and system capabilities, as well as onboard storage and energy resources sufficiency.	Command validation tools, resource allocation models, payload monitoring systems.	Safe and efficient execution of payload missions with minimized risk to satellite integrity.
Power supply safety	Monitors key parameters such as current, subsystem current, battery discharge depth, and charging current under various spacecraft working modes. Adjusts control parameters, switches equipment, or turns off payloads to ensure energy safety upon detecting abnormalities.	Power management systems, current and voltage sensors, energy consumption analysis tools.	Continuous operation within safe energy parameters, preventing power-related failures.
Safety mode	When attitude or power supply abnormalities occur that cannot be immediately corrected, it activates and transfers the spacecraft to safety mode for ground processing. This includes stopping all payload missions, shifting to minimum energy mode, emergency or stop control mode, and maintaining a minimal working state.	Attitude control systems, power management protocols, automated safety mode activation.	Preservation of spacecraft integrity during critical failures by reducing operational demands and facilitating recovery.
System recovery	Stores critical parameters in the health data storage unit and subsystems. Allows for quick recovery to the previous working mode after an intelligent control unit reset, switch, or other failure modes by reading stored parameters from subsystems or the health data storage unit.	Data backup and recovery solutions, health data storage units, recovery protocols.	Rapid restoration of spacecraft functionality post-failure, minimizing downtime and mission disruption.
System reconfiguration	Required in cases of equipment failure, working mode errors, or other abnormalities to restore system status. Methods include power cycling, SW reboot, transferring tasks between devices or subsystems, SW maintenance, and component switching.	Reconfiguration algorithms, fault-tolerant systems, component switching mechanisms.	Adaptability and resilience in maintaining operational status through dynamic system adjustments.
Anomaly response	Identifies and responds to anomalies by isolating affected components or systems and implementing predefined mitigation strategies to prevent further damage or system degradation.	Anomaly detection SW, isolation protocols, mitigation strategies.	Quick isolation and mitigation of anomalies to maintain system integrity and prevent cascading failures.

© SAE International

(Continued)

Table 5.2 **(Continued)** Health management strategies in spacecraft systems.

Strategy	Strategy description	Implementation tools/ techniques	Expected outcome
Environmental adaptation	Adjusts spacecraft operations based on environmental conditions and external threats, such as solar flares or cosmic radiation, to protect sensitive components and ensure mission continuity.	Environmental sensors, adaptive control systems, radiation shielding techniques.	Enhanced resilience and adaptability to space environment changes, ensuring long-term mission success.
Data integrity assurance	Ensures the accuracy, reliability, and timeliness of health data through redundancy, error checking, and secure data transmission protocols.	Redundant systems, EDAC codes, secure communication protocols.	High integrity and reliability of health data for accurate system monitoring and decision-making.
Mission continuity planning	Develops and implements strategies to ensure mission continuity in the face of system failures or anomalies, including backup systems activation and alternative mission planning.	Backup systems, contingency planning tools, mission redefinition strategies.	Ensured mission objectives achievement despite unforeseen system failures or operational challenges.

5.2.1.
Challenges

Managing the health of spacecraft involves addressing several significant challenges, largely because most monitoring systems rely heavily on expert knowledge to develop and maintain. These systems use predefined limits to monitor telemetry values, known as the "out-of-limits" (OOL) technique, which requires a continuous analysis of telemetry data and imposes a significant workload on space operations engineers [5.1]. As systems become more complex, such traditional methods struggle to keep up with the increasing operational demands. There are also other challenges:

- Complex internal structure: The intricate internal structures of spacecraft systems make it difficult to establish precise failure mechanism models. These complex architectures involve many interconnected components, which require detailed and sophisticated modeling to predict and diagnose failures accurately.

- Large data volume: Spacecraft generates vast amounts of telemetry data, primarily representing normal operation conditions. This immense data volume makes identifying subtle variations and life distribution patterns that indicate potential issues challenging. Analyzing these data effectively requires advanced algorithms and substantial computational resources.

- Redundancy and working modes: Spacecraft systems are often designed with many redundant components to enhance reliability. Each of these components can operate in different normal working modes, further complicating the evaluation of the overall health status. Differentiating between normal operational variations and actual faults is critical for accurate health assessment.

- Harsh operating environment: The unpredictable and harsh space environment adds to the complexity of health management. Factors such as extreme temperatures, radiation, and microgravity can affect the performance and longevity of spacecraft components, making it essential to have robust health management systems that can adapt to these conditions.

5.3.
FDIR

The limitations of traditional monitoring methods have led to the exploration of innovative and advanced monitoring systems. FDIR systems are critical components of spacecraft health management as they ensure the continuous operation of spacecraft by identifying, diagnosing, and responding to faults in the system. These methods aim to automate the detection and prediction of system behaviors to reduce the workload by [5.2]:

- Early identification of anomalies and unexpected behavior.
- Predictive health monitoring.
- Reduced dependency on ground operations.

FDIR systems are integral to spacecraft health management and, when combined with advanced technologies and methodologies, significantly enhance the reliability and longevity of spacecraft operations.

5.3.1.
Fault Detection

We can detect system faults in spacecraft manually or automatically, depending on the operating modes and the urgency of restoring the system. Manual detection relies on human senses to identify issues quickly, such as the obvious failure of a light switch when the light does not turn on. The advantage of this method is that it incurs no additional costs from complex system designs.

5.3.1.1.
Built-In Testing (BIT)

A more sophisticated method for fault detection and isolation is BIT, which does not require external test equipment. BIT can range from simple indicators, like a light that turns on when equipment fails, to advanced systems where a resident computer

generates test signals and evaluates system responses. Users can continuously operate BIT, interleaving it with other operations, or initiate it on command. For example, during power-on self-tests, the system runs diagnostics using HW sensors and SW error-correcting codes. The specifics of BIT implementation, determined by the system designer, can include:

1. Additional HW: BIT often requires extra HW beyond that needed for the primary function, affecting reliability and cost. Designers must balance these factors for an effective solution.

2. Fail-safe design: BIT circuitry should be fail-safe, meaning its failure should not affect system performance. To prevent impairment, we should isolate the inputs and outputs of BIT from normal channels.

3. Test points and self-test meters: BIT may include features like test points and self-test meters to help technicians quickly identify faulty components, thus reducing mean time to repair (MTTR).

5.3.1.2.
Centralized vs. Decentralized Architecture

In a centralized approach, a single unit monitors and reports system performance issues. This unit collects data from lower levels, determines if a failure has occurred, and reports faults. This passive BIT method does not use test patterns, so it may not fully monitor the system. Active BIT, which writes and compares test patterns, can be more comprehensive but requires system operation interruption.

- Passive BIT: Monitors system performance without a test pattern generator.
- Active BIT: Writes test patterns to units and compares results, useful when modules are not in use (interleaving BIT).

Decentralized BIT places detection capabilities at the maintainable unit level. Each unit detects its own failures and reports them to a higher level. Both passive and active BIT can be used, and modules can be taken offline for comprehensive testing without disrupting overall system operations.

- Self-test capability: Each unit can test itself.
- Isolation: Each unit's BIT functions independently of other system data.
- Synchronization: System functions ensure synchronization with BIT operations.

5.3.1.3.
Voting Scheme

A voting scheme is another effective fault detection technique. It involves processing data through three or more redundant computers and declaring a failure if one unit's output differs from the others. This method, commonly used in decentralized architectures, requires additional resources but is highly effective. For example, the Orbiter's inertial measurement system uses three redundant units to compare real-time data and achieve majority agreement.

5.3.2.
Fault Isolation

Once a fault is detected, the next crucial step is to actively locate the cause of that fault. This process, known as fault isolation, must consider several factors, including system complexity, the urgency of the turnaround time, the repair location, and the skill level of the personnel involved.

In many cases, technicians carry out manual fault isolation. This process involves visually inspecting for obvious signs of failure, such as burned-out components, or using external test equipment to diagnose the system. This method

can be effective for simpler systems or when the fault is easily observable. For example, a technician might visually inspect circuit boards for burned-out components or use a multimeter to check for electrical continuity. However, manually performing fault isolation can consume a lot of time and requires skilled technicians to accurately identify the problem. This is particularly challenging in space missions, where maintenance engineers must manually inspect the system, leading to high costs and logistical challenges. The need for specialized tools and expertise further adds to the expense and complexity of manual fault isolation in space environments.

- BIT systems: These continuously monitor system performance and can automatically isolate faults in specific components or regions. Diagnostic algorithms achieve this by analyzing system behavior and identifying anomalies. BIT systems can operate in real time, providing immediate feedback and reducing the need for manual intervention.

  BIT can sometimes isolate faults at a specific system level or region. For example, in a series of connected components, BIT might identify the faulty region, but not the exact component. This limitation requires additional manual inspection or more sophisticated testing methods to pinpoint the exact failure.

- Voting scheme: A voting scheme, which involves redundant systems comparing outputs, can more accurately identify the failed unit. When one unit's output differs from the others, the system isolates the faulty unit. However, if two units indicate a fault, identifying the exact failure becomes challenging because of the lack of a majority reference.

5.3.2.1.
Automated Techniques

Automated fault isolation significantly reduces the time and cost associated with manual inspections. By quickly identifying and isolating faults, automated systems minimize downtime and improve overall system reliability. This is particularly important in space missions, where maintenance windows are limited, and the cost of manual inspections is prohibitively high. Several advanced automated techniques have been developed to enhance fault isolation in complex systems.

- Model-based diagnosis: This technique uses mathematical models of the system to predict and diagnose faults. By comparing real-time data with model predictions, we can identify discrepancies and use them to isolate faults. Model-based diagnosis is particularly useful in systems with complex interactions and dependencies.

- AI-based diagnosis: Data-driven algorithms can analyze large datasets to detect patterns and anomalies that may indicate faults. These techniques can improve fault isolation accuracy and adapt to changing system conditions. Neural networks can learn to recognize fault signatures and provide real-time diagnostics [5.3].

- Redundancy and voting schemes: Redundant systems with multiple identical components can use voting schemes to isolate faults. When one component's output differs from the others, the system flags it as faulty. Critical systems often use this highly reliable method where failure is not an option.

- Remote sensing and diagnostics: In space missions, remote sensing technologies can be used to monitor system health and diagnose faults from Earth. This approach reduces the need for on-site inspections and allows for continuous monitoring of spacecraft systems.

- Boundary scan technique: The boundary scan technique, commonly used in integrated circuit (IC) testing, allows for detailed fault isolation by dividing the IC into regions accessible via scan operations. This technique overcomes test access problems and provides precise fault isolation, essential for complex systems.

5.3.2.2.
Boundary Scan Technique

The boundary scan technique is a recent and advanced method in the IC industry, offering significant advantages for fault isolation, especially in complex systems such as spacecraft. This technique divides the IC into regions accessible via scan operations, providing a detailed and efficient way to test and isolate faults. The connection of each IC function pin to a boundary scan cell (BSC) creates a series of interconnected BSCs from the test data input pin to the test data output pin. This allows for serial access and precise fault detection.

How boundary scan works:

1. Normal operation: During regular IC operation, signals pass through the BSCs without interference. This seamless integration ensures that the boundary scan circuitry does not disrupt the normal functionality of the IC.

2. Test mode: When the IC enters the test mode, the boundary scan technique becomes active. Applying test stimuli through the BSCs, we capture the resulting signals at the end of the scan path. This process allows for a thorough examination of the IC's performance and helps in identifying faults with high precision.

This technique is particularly useful in spacecraft systems, where traditional access methods may be inadequate because of the system's complexity and harsh operating environments. Organizations can combine detailed unit-level testing enabled by boundary scan for comprehensive system-level verification. It enables remote diagnostics, crucial for space missions where physical access is impossible, allowing engineers to test and diagnose ICs from Earth, ensuring continuous monitoring and maintenance of spacecraft health. The adaptability of boundary scan to various complex systems within spacecraft, such as navigation, communication, and control systems, makes it an ideal solution for the intricate components of modern spacecraft. Additionally, by reducing the need for manual inspections and interventions, boundary scan lowers maintenance costs, which is critical in the high-budget environment of space missions.

This integration can also enable real-time monitoring and diagnostics, immediate fault detection, and isolation in onboard systems. Engineers can transmit data from boundary scan operations back to Earth for analysis, which helps them make informed decisions about maintenance and repairs. Overall, the boundary scan technique significantly enhances the reliability and operational integrity of spacecraft, ensuring mission success and extending the lifespan of space missions.

5.3.3.
Fault Recovery

For a system to recover from a failure, whether manually or automatically, the modes of operation must be predefined and planned based on the types of failures anticipated. The design phase involves identifying critical functions, redundancy levels, and functional paths to

ensure effective recovery. Generally, recovery can be categorized into three types:

1. One hundred percent functional recovery using redundant system components.
2. Functional recovery using an alternative path.
3. Degraded functional recovery.

In category (1), the system reports the failure when a component fails, and it enables the activation of a redundant or backup unit, either manually or automatically. In autonomous systems such as satellites, these systems must continue operating independently because of the impracticality of real-time human intervention in space.

Constraints on weight, space, and cost inherently limit resources on space missions. Therefore, instead of having a redundant string or unit for every critical function, systems often employ an alternative path for recovery, as described in category (2). This alternative path may not fully replicate the original function's capabilities but can provide a viable solution within the constraints. For instance, in a space station, if the cooling loop of the thermal control system fails and cannot cool the electronics on the cold plates, the crew could redirect cool air from the environmental control and life support system to prevent overheating. This method ensures continued operation, albeit at reduced efficiency, by rerouting available resources.

In the worst-case scenario, where redundant systems are unavailable, as in category (3), the system must operate at a minimal capacity to ensure the safety of the crew and the spacecraft. We evaluate critical functions to determine which components we can temporarily shut down without compromising the spacecraft's control until we can make repairs. For instance, if the solar array panels sustain damage and cannot generate enough electrical power, the crew and spacecraft assess critical functions to determine which components they can temporarily deactivate without compromising the spacecraft's control until repairs can be made. Even critical systems might need to operate at a compromised level to maintain overall functionality and protect the mission.

By categorizing and planning recovery methods during the design phase, space missions can ensure robust fault tolerance and resilience, maintaining operational integrity in the face of failures. Adopting this proactive approach is crucial for the success and safety of space missions, given the unforgiving environment and limited real-time support from Earth.

5.3.4.
Hierarchical FDIR Architecture

Implementing FDIR systems also involves conducting failure mode and effects analysis (FMEA) at both system and subsystem levels. This analysis helps identify potential failure modes, their effects on the system, and the recovery actions. For example, the ESEO spacecraft utilized commercial off-the-shelf (COTS) electronic components and FMEA analysis to ensure mission success while achieving cost savings. This helps to form a hierarchical FDIR architecture, structured to confine failures to the lowest possible levels to minimize system outages and ensure high availability:

- Unit level: Detects and isolates faults within individual units, such as sensors or actuators.

- Subsystem level: Manages faults within larger subsystems, integrating multiple units.

- System level: Oversees the entire spacecraft's health, coordinating recovery actions across all subsystems.

Realizing the above includes the implementation of FDIR systems, which also involve HW and SW aspects. On the SW side, FDIR systems typically aim to isolate and recover faults at different levels, such as unit, subsystem, or equipment levels. The deterministic approach relies on predefined tables containing selected monitoring items and corresponding recovery actions. Designers base these tables on historical data and expert knowledge, and then implement them in the avionic SW (ASW).

However, some challenges for the traditional methods remain even when implementing FDIR systems, such as the need to accommodate varying telemetry inputs and fault monitor logic specific to each spacecraft. This requires a customized solution for each mission, leading to high design, development, coding, validation, and testing costs. The process of recovering these costs over multiple spacecraft is challenging, resulting in high expenses and time-consuming development, coding, validation, and testing.

More recent advancements include the use of cognitive automation that enables context-sensitive reactions to unexpected failures by utilizing comprehensive knowledge about the system's state, operational capabilities, and the impact of faults and recovery actions. This approach is particularly effective in reducing the onboard reaction time and increasing the spacecraft's resilience to failures.

AI techniques are also increasingly being integrated into FDIR systems to enhance their capabilities. Examples of AI-based FDIR include:

- Support vector machines: Used for fault diagnosis and anomaly detection in spacecraft telemetry.

- Random forest: Applied for condition monitoring and early failure detection.

- Neural networks: Employed for active fault-tolerant control, improving reaction times and success rates.

But implementing these systems onboard spacecraft also presents several challenges:

- Computational power: Traditional onboard computers (OBCs) may not have the computational power to run complex ML models.

- Data availability: High-fidelity synthetic or simulated data are often limited, making it difficult to train accurate models.

- Memory constraints: ML models may exceed the memory and execution time budgets of OBCs.

Recent advancements in space processor performance and dedicated vector acceleration have made it possible to use new OBCs with enhanced capabilities. These improvements enable the deployment of AI-based FDIR systems, targeting medium to high-dependability applications without requiring ML inference to be supervised by a fault-tolerant engine.

5.4.
Toward Integrated Health Monitoring

The traditional approach, while foundational in the aerospace industry, exhibits several limitations that necessitate a transition toward more modern methodologies. This is because it often operates reactively, addressing issues as they arise rather than proactively mitigating them, potentially leading to increased downtime and jeopardizing mission success. They also underutilize the extensive data generated by an asset, missing opportunities for optimization and oversight. However, a significant drawback comes from the reliance on predefined thresholds to trigger alerts or actions, which may not adequately account for the unique conditions of each mission or the gradual wear and tear of components. Many of these approaches lack the predictive capabilities necessary for managing unknown issues, a necessity for extended space missions and deep space exploration. Another limitation arises from some level of manual intervention required for problem resolution. Coupled with high maintenance and operational costs because of the need for redundancy and backup systems, it underscores the need for a more adaptable, resource-optimized framework that can make long-duration and deep space missions more capable. Figure 5.2 provides a more comprehensive representation of broader concepts and activities for spacecraft health management. This is useful for understanding the multifaceted approach required to ensure the operational integrity and safety of spacecraft.

The following discussion focuses on condition-based maintenance (CBM), predictive health management, and fault-tolerant design. The discussion of the rest of the health management concepts highlighted in Figure 5.2 is provided in Chapter 10.

Figure 5.2 Health management concepts for spacecraft.

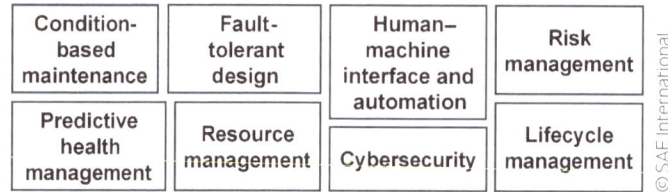

Condition-based maintenance	Fault-tolerant design	Human–machine interface and automation	Risk management
Predictive health management	Resource management	Cybersecurity	Lifecycle management

© SAE International

5.4.1.
CBM

CBM is a proactive method that involves monitoring the current condition of spacecraft components and performing maintenance accordingly. This method contrasts with traditional maintenance schedules that are time-based, regardless of the component's condition. The operations of the ISS provide a real-world example of this, where a sophisticated network of sensors continuously monitors various systems, from life support to power generation. These sensors provide real-time data on the condition of each component, allowing the crew and ground control to identify potential issues before they become critical. For instance, if a sensor detects an anomaly in a solar panel's output, the crew and ground control can schedule maintenance immediately to address the issue, rather than waiting for a scheduled check [5.4].

Without such measures, spacecraft risk unexpected failures, which can be catastrophic in space. For example, in 1997, the Mir space station experienced the failure of the oxygen generation system because of unexpected component failure, which more advanced condition monitoring might have prevented.

Sensing technology: Health and usage monitoring systems (HUMSs) are essential for ensuring the ongoing health and optimal performance of various systems. HUMS involves the continuous monitoring of spacecraft components using advanced sensors and diagnostic tools, detecting anomalies, and predicting potential failures before they occur. HUMS relies on a variety of sensors to collect data on different aspects of the spacecraft's health, including vibration, thermal, strain gauge, acoustic sensors, etc. The system preprocesses these raw data to filter and amplify them before any onboard processing.

There are several implementation challenges and considerations with the technology. First, the miniaturization and power efficiency of sensors and acquisition systems are crucial because of the limited space and power resources available on spacecraft. Second, these components must be robust enough to withstand the harsh conditions of space, including extreme temperatures, radiation, and vacuum. Efficient data management also poses a significant challenge, as handling and transmitting large volumes of sensor data, particularly for missions deep in space, requires sophisticated solutions. Lastly, the reliability and redundancy of HUMS components are critical. High reliability ensures consistent performance, while built-in redundancy safeguards against system failures, ensuring that monitoring continues even if one part of the system fails.

Some of the key concepts include the following:

- Onboard processing: With advancements in computing, spacecraft started incorporating onboard processors capable of analyzing data in real time. This reduced the dependence on ground stations and allowed for quicker anomaly detection and response.

- Model-based systems: Instead of relying solely on predefined thresholds, modern spacecraft use model-based systems that predict the expected behavior of the spacecraft based on its current state and inputs. Deviations from this model can show potential issues.

- ML: Recent advancements have seen the integration of ML and AI techniques for predictive maintenance. These methods can detect subtle patterns in the data, predict potential failures, and even suggest corrective actions.

- Integrated systems: Modern spacecraft health monitoring is not just about tracking the health of individual components but understanding the integrated health of the entire system. This holistic approach ensures that we consider the interactions between different systems, which leads to a more comprehensive understanding of the spacecraft's health.

As commercial space travel becomes more prevalent, people expect CBM to play an essential role in ensuring the safety and comfort of passengers. Monitoring systems will track life support systems, cabin pressure, and other environmental conditions critical for human habitation. It can also be used to self-diagnose issues and perform self-repairs or adjustments without waiting for instructions from Earth.

5.4.2.
Predictive Health Management

Predictive health management plays an important role in correlating health information from various subsystems into a comprehensive assessment of the asset's capabilities. This system predicts potential losses in asset capability, coordinates recovery strategies with mission managers, and provides situational awareness to both crew and ground personnel. To enhance the safety and reliability of space exploration, real-time capability is necessary for quickly predicting and detecting critical failures, providing enough time for successful recovery measures. Such predictive technologies, including fault detection, isolation, and prognostic methods, along with intelligent or expert systems, are integral to an overall failure identification and recovery solution.

Historically, GNC flight SW has performed various FDIR functions to ensure mission safety in the face of system failures. However, responses are at the subsystem level, with a limited examination of system-level issues. Predictive health technologies enable enhanced capabilities where vehicles and crew are more isolated from ground support. For instance, an Autonomous Flight Manager's role includes planning new flight paths in response to critical failures and coordinating with the crew survival system to assess trajectory alterations or activate abort sequences. Most of the functions will be largely automated and will require minimal human input. System analysis and trade studies are necessary to identify specific failure modes and risks that need mitigation.

Ideally, each subsystem should diagnose and predict its health status. This approach enables collaborative decision-making for a comprehensive analysis at the vehicle system level.

Ideally, each subsystem should handle its own fault detection and isolation, as well as collect, store, and transmit health assessment data for operations and maintenance, and send them to ground control as needed. At the system level, the health data from various subsystems can be cross-correlated to assess the overall condition and capabilities of the spacecraft. Key subsystems identified for this purpose include the following:

- Propulsion: The propulsion subsystem's health management is vital for mission success, especially for long-duration missions. It includes monitoring for propellant leaks and ensuring the reliability of propulsion activities in space. Technologies such as leak detection systems, which use networks of lightweight sensors, are crucial for monitoring critical elements such as helium and hydrogen.

- Structure and thermal protection system (TPS): The structural integrity of spacecraft is paramount for safety. Advanced sensor technologies, such as fiber Bragg grating and acoustic emission sensors, are used to monitor the structural health of spacecraft components in real time. This is particularly important in light of accidents like the Columbia disaster, where structural failure had catastrophic consequences.

- Power and actuators: These systems are integral to the operation of other subsystems. Faults in power and actuators can lead to system-wide issues. Health management in this area focuses on detecting and isolating faults that could disrupt the operation of the entire system.

- Avionics and GNC: Avionics systems, encompassing flight SW, sensors, and operational processes, interact with other subsystems and contribute to all aspects of vehicle operations. Avionics systems require health management technologies to enhance capabilities, including predictive aspects for FDIR.

Figure 5.3 demonstrates how health management integrates across various spacecraft subsystems to form a comprehensive solution. This integrated approach significantly enhances situational awareness and decision-making capabilities for both crew and ground personnel. It enables the prediction of potential losses in vehicle capability and effectively coordinates recovery strategies. Developing these health management systems requires a deep understanding of the latest advancements in each subsystem's health management. While each subsystem has its unique aspects, shared technologies exist, which can improve risk mitigation across different systems.

We target these technologies to yield the highest return on investment and speed up technological development. For instance, propulsion health management is crucial for subsystems that are critical yet hard to access. With structures and TPS, predicting structural integrity before failures occur is vital. Similarly, detecting and isolating faults in power and actuator systems is essential for maintaining the overall functionality of the spacecraft.

The health management system in space adopts a hierarchical and distributed structure, which streamlines data processing and facilitates quick, system-level decision-making. This system can diagnose faults, manage fault recovery, and support maintenance decisions. Different subsystems have varied health management needs. For example, mechanical devices use fault prediction based on performance degradation, while electronic devices rely on built-in tests for

Figure 5.3 An integrated health management system.

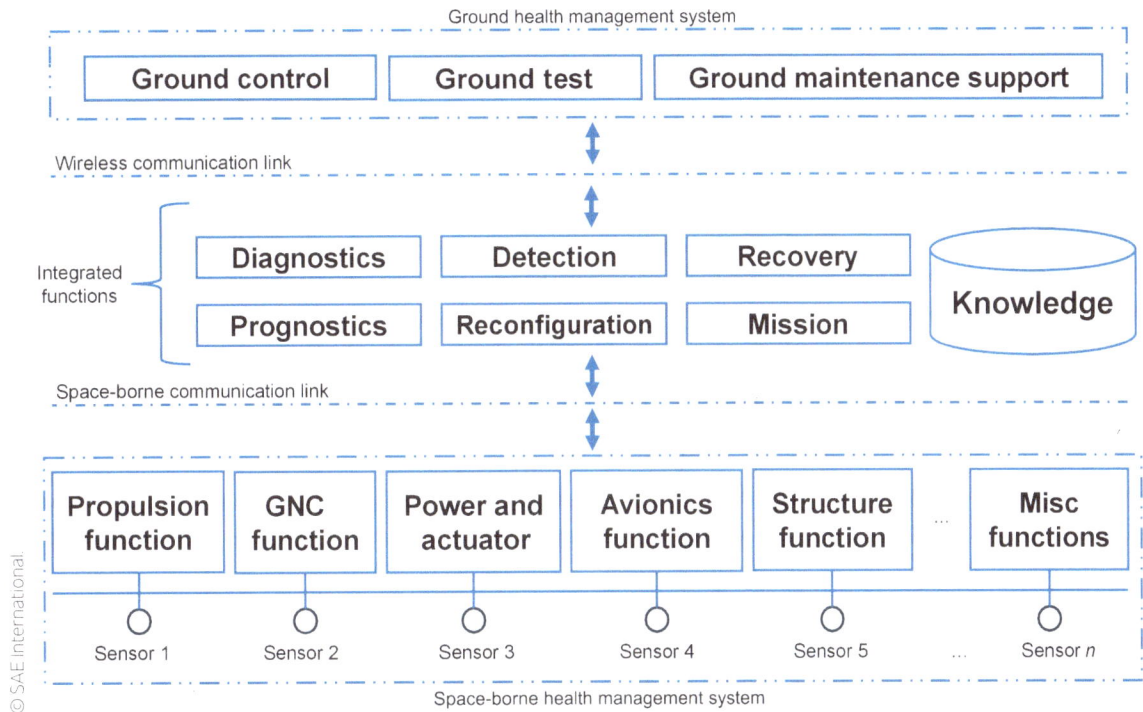

early failure detection. Specific subsystems, like the attitude control system, use advanced sensors, including optical fibers, for enhanced monitoring. Being proactive in multiple subsystems makes the spacecraft more reliable. By analyzing data from sensors and operational parameters, we can gain insights into the health status and make predictions for the engine and other components. Data fusion from various sources, like engine sensors and maintenance history, enables fault detection and maintenance decision-making. The power system's health, especially the battery's performance, is critical for spacecraft operation. Monitoring electrical signals and environmental changes helps evaluate the system's health and predict the battery's lifespan.

However, because of limited storage and computing power on the spacecraft, extensive historical data storage and processing for complex fault diagnosis or long-term prediction are challenging.

Therefore, the ground health management system plays a crucial role in retrieving health status data from the onboard system. It uses its storage and computational capabilities for analysis, identifies abnormalities, and makes maintenance decisions, often with manual intervention.

This integrated health management approach, combining onboard monitoring with ground-based analysis, is essential for ensuring the safety and success of space missions as it enables a comprehensive assessment of spacecraft health, proactive maintenance, and enhancing the reliability of space missions.

5.4.3.
Fault-Tolerant Design

Whenever repair options are limited and system failures can have critical consequences, having a fault-tolerant design becomes important. Because of the unique and challenging environment of space, several concepts become focal points of discussion:

- Harsh space environment: Spacecraft operates in extreme conditions, including high radiation, vacuum, and significant temperature fluctuations [5.5]. These conditions can cause unexpected failures in spacecraft systems. Since the success of space missions, whether it is for research, communication, or exploration, the continuous functioning of the spacecraft has become paramount. The fault-tolerant design ensures that these systems can continue to operate even when some components fail.

- Limited maintenance options: Once the spacecraft is launched, there are extremely limited opportunities for any physical repairs, especially for unmanned missions. But even with crewed missions, the safety of the astronauts depends on the reliability of the spacecraft systems [5.6]. Fault-tolerant designs are essential to manage and mitigate failures during navigation and communication so that the asset remains operational even in the event of component failures.

- Cost-effectiveness: Space missions are expensive. A fault-tolerant design increases the reliability and longevity of the space-craft, ensuring that the investment yields the maximum return in terms of data and mission objectives achieved. A key enabler is

redundancy or backup components that can affect the cost of systems.

- Autonomous operation: Modern spacecraft are often equipped with systems that allow them to detect and diagnose problems, and then take corrective action without human intervention. This autonomous operation is a part of a fault-tolerant design and is important for missions far from Earth.

An example of fault tolerance is the design of the HST, which incorporated redundant systems, enabling it to continue functioning despite various failures. For instance, when one of its gyroscopes failed, the telescope could still function with the remaining gyroscopes. NASA conducted multiple SMs to repair and upgrade the HST, highlighting the significance of fault tolerance in extending mission life.

The strategy is to use a layered approach designed to ensure the spacecraft's operational integrity and safety. This strategy encompasses various components, each tailored to address specific aspects of spacecraft operation and potential anomalies. These components include system monitoring, payload mission safety, power supply security, transitioning to a safe mode, system recovery, and reconfiguration strategies [5.7]. The successful execution of these strategies relies on a foundation of advanced technologies and methodologies, such as managing telemetry and telecommand anomalies. Figure 5.3 illustrates the intricate structure of the spacecraft system layer health management strategy, highlighting its multifaceted approach.

- System monitoring strategy: This involves continuous surveillance of the spacecraft's operational parameters to detect deviations from expected performance. Advanced sensors and diagnostic algorithms play a crucial role, providing real-time data that can preemptively identify issues before they escalate.

- Payload mission safety strategy: Tailored to protect the spacecraft's mission-critical components, this strategy ensures that the payload remains functional and secure, even in the face of system-wide anomalies. It involves redundancy designs and isolation mechanisms that safeguard mission-critical data and functionality.

- Power supply safety strategy: Given the critical importance of power in maintaining spacecraft operations, this strategy focuses on monitoring and managing the power supply system to prevent failures. It includes redundancy in power sources, real-time power consumption analysis, and emergency power rationing protocols.

- Transfer to safety mode strategy: In the event of a significant anomaly, this strategy involves transitioning the spacecraft to a predefined safe mode. This mode minimizes operational risks by reducing system functionality to essential services only, preserving system integrity, and allowing for detailed fault analysis and recovery efforts.

- System recovery strategy: This component outlines the steps for restoring the spacecraft to its full operational capacity following an anomaly, guaranteeing a controlled return to normal operations.

- System reconfiguration strategy: To adapt to changing conditions or recover from faults, this strategy enables the dynamic reconfiguration of system components. It involves SW and HW flexibility, allowing the spacecraft to change its operational setup based on current needs and constraints.

To implement each health management strategy, a suite of basic technologies enables sophisticated fault detection, accurate diagnosis, and efficient anomaly resolution. These technologies include data fusion techniques that integrate data from multiple sources for a comprehensive understanding of system health, and SW development practices that ensure the reliability and adaptability of health management systems. The anomaly event management provides a standardized approach to managing unexpected events, ensuring that the spacecraft can maintain its mission objectives even in the face of unforeseen challenges.

5.5.
Sensor Placement Optimization

Proper sensor placement ensures comprehensive monitoring, improves fault detection and isolation, and enhances the overall reliability of spacecraft systems [5.8]. The primary objectives of sensor placement optimization are as follows:

- Maximize coverage of critical components and areas.

- Minimize redundancy while ensuring fault tolerance.

- Optimize sensor performance by placing them in optimal operational environments. Minimize interference from other spacecraft systems.

- Ensure ease of maintenance and accessibility.

5.5.1.
Methods and Techniques
FEA: FEA is a computational technique used to predict how a structure will react to real-world forces, vibration, heat, fluid flow, and other physical effects. FEA breaks down a large,

complex system into smaller, simpler parts called finite elements. Analysts break down these elements and analyze them, and then combine the results to provide a comprehensive picture of how the structure behaves under various conditions.

FEA enables the identification of critical stress points in the spacecraft's structure where sensors should be placed to monitor stress, strain, and potential failures, for example, during launch, the spacecraft experiences significant vibrations and forces. FEA can simulate these conditions to find the best locations for placing accelerometers and strain gauges.

Steps include:

- Create a detailed model of the spacecraft structure.
- Apply expected loads and conditions (e.g., vibrations, thermal changes).
- Analyze the model to identify areas of high stress or strain.
- Place sensors at these critical points to monitor structural health.

Genetic algorithms (GAs): GAs are inspired by the process of natural selection and evolution. They are used to solving optimization problems by simulating the process of natural evolution. GAs involve a population of candidate solutions that evolve over generations. GAs can optimize sensor placement by considering multiple objectives, such as maximizing coverage, minimizing redundancy, and ensuring sensor performance; for example, it can help determine the optimal placement of temperature sensors to monitor the thermal control system effectively.

Steps include:

- Initialize a population of possible sensor placements.
- Evaluate the fitness of each placement based on predefined criteria (e.g., coverage, performance).
- Select the best-performing placements to create the next generation.
- Apply crossover and mutation to introduce variations.
- Repeat the process for several generations until an optimal placement is found.

Particle swarm optimization (PSO): PSO is inspired by the social behavior of birds flocking or fish schooling. It is a population-based optimization technique. In PSO, each candidate solution is called a particle. These particles move through the solution space, influenced by their own best-known position and the best-known positions of other particles. PSO can optimize sensor placement by finding the best positions for sensors to achieve objectives, such as high coverage and minimal interference; for example, PSO can optimize the placement of radiation sensors on the spacecraft's outer shell to ensure comprehensive environmental monitoring.

Steps include:

- Initialize a swarm of particles with random positions.
- Evaluate the fitness of each particle based on objectives (e.g., coverage, interference).
- Update each particle's velocity and position based on its best-known position and the swarm's best-known position.
- Repeat the process until convergence to an optimal placement.

Simulated annealing (SA): SA is inspired by the annealing process in metallurgy, where a material is heated and then slowly cooled to remove defects. SA is a probabilistic optimization technique that explores the solution space by accepting both improvements and occasional worse solutions to escape local optima. SA can optimize sensor placement by balancing trade-offs between coverage, performance, and cost; for example, SA can optimize the placement of pressure sensors within the propulsion system to monitor performance effectively.

Steps include:

- Initialize a random sensor placement configuration.
- Evaluate the fitness of the configuration based on objectives (e.g., coverage, cost).
- Generate a new configuration by making small changes to the current one.

- Accept the new configuration if it improves the objective or, with a certain probability, if it is worse.
- Gradually reduce the probability of accepting worse solutions (cooling schedule).
- Repeat the process until convergence to an optimal placement.

5.5.2.
Examples

Sensor placement optimization is essential for effective spacecraft monitoring and management. By leveraging advanced optimization algorithms and methodologies, researchers can develop sensor placement strategies that maximize coverage, minimize redundancy, and enhance the overall reliability of spacecraft systems (Table 5.3).

Table 5.3 Reliability of spacecraft systems.

Case study	Scenario	Methods	Outcome	Reference
Structural health monitoring	Monitoring the structural integrity of the spacecraft during launch, orbit, and reentry.	Using FEA to identify critical stress points and placing strain gauges and accelerometers at these points.	Enhanced ability to detect early signs of stress or damage, leading to timely maintenance and repairs.	[5.8]
Thermal control	Ensuring the spacecraft maintains optimal temperature ranges.	Placing temperature sensors at key locations where heat generation and dissipation occur, optimized using GA.	Effective thermal management, preventing overheating or freezing of critical components.	[5.5]
Propulsion system monitoring	Monitoring the health and performance of propulsion systems.	Strategic placement of pressure and temperature sensors within the propulsion system, optimized using PSO.	Real-time monitoring of fuel lines and combustion chambers, ensuring safe and efficient propulsion.	[5.6]
Environmental monitoring	Measuring external conditions such as radiation levels and micrometeoroid impacts.	Deploying radiation sensors and impact detectors on the spacecraft's outer shell, optimized through SA.	Comprehensive environmental monitoring, enhancing the spacecraft's ability to withstand and respond to external hazards.	[5.7]

5.6.
Summary

Space missions demand a lot of investment and precision. This chapter discusses these requirements with reliance on telemetry data and predefined limits for advanced predictive health management systems. Modern technologies promise to minimize dependency on ground-based analysis and allow for a quicker response to anomalies. This movement toward a unified, predictive approach underlines the importance of continuous monitoring, sophisticated diagnostics, and autonomous correction in maintaining spacecraft health. It represents a forward-thinking strategy to ensure mission success amid growing complexities, highlighting the critical role of advanced technologies in the future of space exploration and the economic implications of such advancements in the aerospace sector.

References

5.1. Saleh, J.H. and Marais, K., "Flexibility in System Design and Implications for Aerospace Systems," *Acta Astronautica* 59, no. 8-11 (2006): 833-841, doi:https://doi.org/10.1016/j.actaastro.2006.02.002.

5.2. National Aeronautics and Space Administration, "NASA Technology Roadmaps: TA 11, Modeling, Simulation, Information Technology & Processing," NASA, 2011.

5.3. Yairi, T., Takeishi, N., Oda, T., Nakajima, Y. et al., "A Data-Driven Health Monitoring Method for Satellite Housekeeping Data Based on Probabilistic Clustering and Dimensionality Reduction," *IEEE Transactions on Aerospace and Electronic Systems* 53, no. 3 (2017): 1384-1401.

5.4. Stamatelatos, M. and Dezfuli, H., *Probabilistic Risk Assessment Procedures Guide for NASA Managers and Practitioners*, 2nd ed. (Washington, DC: National Aeronautics and Space Administration, 2011).

5.5. Sun, Y., Zhang, C., Ji, H., and Qiu, J., "A Temperature Field Reconstruction Method for Spacecraft Leading Edge Structure with Optimized Sensor Array," *Journal of Intelligent Material Systems and Structures* 32, no. 17 (2021): 2024-2038.

5.6. Balakrishnan, N., Devasigamani, A.I., Anupama, K.R., and Sharma, N., "Aero-Engine Health Monitoring with Real Flight Data Using Whale Optimization Algorithm Based Artificial Neural Network Technique," *Optical Memory and Neural Networks* 30 (2021): 80-96.

5.7. Paek, S.W., Kim, S., and de Weck, O. (2019). Optimization of Reconfigurable Satellite Constellations Using Simulated Annealing and Genetic Algorithm. *Sensors*, 19(4), 765.

5.8. Ostachowicz, W., Soman, R., and Malinowski, P., "Optimization of Sensor Placement for Structural Health Monitoring: A Review," *Structural Health Monitoring* 18, no. 3 (2019): 963-988.

In 1945, Arthur C. Clarke introduced the concept of geostationary satellites in his paper titled "Extra-Terrestrial Relays: Can Rocket Stations Provide Global Radio Coverage?" published in *Wireless World* magazine. His proposition detailed the establishment of satellites in a fixed orbit, approximately 35,786 km above Earth, to facilitate uninterrupted telecommunications, weather monitoring, and broadcasting services. Clarke's idea charted the course for future satellite communications, heralding an era of global connectivity long before the technology to launch satellites even existed. It started taking shape with the launch of Syncom 2 by NASA in 1963, marking the world's first foray into geostationary communications satellites. This step followed closely after the deployment of Early Bird (INTELSAT I) in 1965. Early Bird was the first commercial satellite to occupy the geostationary orbit envisioned by Clarke, marking the start of the global satellite communications era.

As the communications industry advanced, the importance of telemetry became increasingly clear. Telemetry enabled the continuous monitoring of spacecraft equipment, affirming its performance under the harsh conditions of launch and space operations. Its utility spanned the entire lifecycle of space missions, from the manufacturing process and testing of rockets and payloads to the critical moments of launch. However, the aerospace industry faced challenges in fully using telemetry's potential. Some manufacturers may not prioritize comprehensive telemetry systems because of various concerns, including cost, complexity, and testing delays, potentially compromising long-term equipment reliability.

In this context, health management technology marked a significant leap forward, offering the potential to improve spacecraft reliability and maintenance significantly. Advances in predictive algorithms and diagnostic technology, coupled with a deeper appreciation of telemetry's value, could revolutionize how the industry approaches spacecraft design, testing, and operation. This chapter explores the role of telemetry in monitoring spacecraft health, tracing its history from early satellite communications to modern applications in predictive maintenance, and detailing how advanced data analysis techniques and ML enhance reliability and performance throughout space missions.

6.1.
Brief History of Telemetry

Originating in 1812 Russia with applications in military operations, telemetry has undergone significant evolution, expanding its reach into meteorology, seismic monitoring, and beyond. The early twentieth century saw its application in pivotal projects such as the Panama Canal construction, leveraging radio meteorological telemetering to overcome environmental challenges. This period marked the beginning of telemetry's integration into broader scientific and engineering disciplines, setting the stage for its critical role in the development of modern aerospace technologies.

The post-World War II era saw significant advancements in telemetry, driven by the emergence of aeronautical electronics and digital computing. Introducing digital computers, exemplified by Louisiana Power and Light's system in 1958, marked a turning point in telemetry application, particularly in missile performance diagnostics and in-flight instrumentation. This period saw telemetry evolve from transmitting a mere 30–40 measurements per flight to more than 600, highlighting the technology's growing complexity and its indispensability for diagnosing and enhancing system performance.

The 1960s represented a golden era for telemetry in aerospace engineering, underscored by the Apollo moon landing and the dawn of unmanned space exploration. Telemetry's transition from analog channels to digital systems facilitated these monumental achievements, enabling precise monitoring and control over space missions. The deployment of satellites, such as ECHO, TIROS I, and TRANSIT, further showcased telemetry's role in fostering global communication, with the establishment of geostationary orbits bringing Clarke's concept to reality.

Despite its advancements, there were various challenges, particularly in standardizing instrumentation for satellite diagnostics due to:

- High dimensionality and volume of data.
- Complex network dependencies between channels.
- Varying sampling frequencies, data gaps, component degradation trends, and operational mode changes.
- Noise and measurement errors from the space environment.

The late 1980s saw a critical shift with INTELSAT's mandate for proper satellite instrumentation, emphasizing telemetry's essential role in quantifying equipment performance. Telemetry networks also faced vulnerabilities, particularly concerning political instabilities in host countries of ground stations. This prompted NASA to explore systems with space-based communication nodes, aiming to enhance performance and reliability. The formation of the Consultative Committee for Space Data Systems (CCSDS) in 1982 was a significant milestone in standardizing space data. It aimed to address challenges related to data reliability within the deep learning (DL) model. This initiative fostered international collaboration, leading to the development of spacecraft "packet telemetry" and "packet telecommand" standards, significantly improving data integrity and the reliability of telemetry data. However, the challenge of balancing system complexity with budgetary and technological constraints persists, necessitating ongoing innovation and standardization of telemetry technologies.

6.1.1.
Using Telemetry in Spacecraft

The journey of spacecraft telemetry began with the launch of Sputnik in 1957. Sputnik introduced a dual-frequency telemetry downlink, which set the stage for future advancements in automated data transmission. Following this milestone, in 1958, the US entered the field with the launch of Explorer 1. NASA's predecessor utilized the DSN as a key signaling station, emphasizing the strategic importance of telemetry in space exploration.

The application of telemetry in the diagnostics and prognostics of space equipment represents a significant shift toward greater mission reliability. These systems allow engineers to perform initial diagnostics, which means they can detect potential equipment failures early. By identifying these issues before they escalate, engineers can prevent the launch of faulty components and reduce the risk of failures. This predictive capability not only extends the operational life of spacecraft but also reduces launch delays and improves satellite service availability.

Monitoring satellite telemetry data for anomalies has been vital for the safe and uninterrupted operation of satellites (Table 6.1). Many scientific papers have discussed its importance and challenges, and engineers often rely on automatic detection systems to identify any deviations from normal operations. However, complex anomalies frequently warranted manual detection, which was error-prone and costly.

Table 6.1 Summary of the challenges in satellite telemetry anomaly detection.

Limitation	Description
Scarcity of anomalies	There are relatively few anomalies in flying missions, making it difficult to evaluate different approaches objectively.
Data collection	No comprehensive data collection from multiple sources, leading to biased evaluations.
Public datasets flaws	Majority of publicly available datasets, benchmarks, metrics, and protocols for time series anomaly detection are flawed and cannot be used for unbiased evaluations.
High dimensionality and volume	Satellite telemetry data involve years of recordings from up to thousands of channels per satellite, creating challenges for analysis.
Complex network dependencies	There are complex dependencies between channels, complicating anomaly detection.
Varying sampling frequencies	Different channels have varying sampling frequencies, leading to data gaps and inconsistencies.
Trends and concept drifts	Data contain trends related to spacecraft component degradation and concept drifts due to different operational modes and mission phases.
Diverse channel types	Channels include a large variety and ranges of physical measures, categorical status flags, counters, and binary telecommands.
Noise and measurement errors	The space environment introduces noise and measurement errors, complicating anomaly detection.
Memory inefficiency	Some algorithms are not optimized for handling large datasets, leading to out-of-memory errors.
Thresholding issues	Some methods use data-specific conditions that may not be optimal for different types of data.
Handling of anomalies in training data	Many algorithms assume there are no anomalies in the training set, which is not true for real-life data.
Single output from models	Some models output predictions for a single channel, making scaling to multiple channels challenging.
GPU support problems	Certain implementations have compatibility issues with newer GPU architectures, limiting computational efficiency.
Real-time and online detection	Many algorithms are not designed for real-time, online, streaming detection.
Modeling multivariate dependencies	Some algorithms do not model dependencies between multiple channels, which is essential for accurate anomaly detection in satellite telemetry.
Handling irregular timestamps	Algorithms often require uniformly sampled time series, while satellite telemetry has irregular timestamps and varying sampling rates.
Learning from rare nominal events	Algorithms struggle to distinguish between rare nominal events and anomalies.

6.1.2.
Ensuring Satellite Reliability

Many satellites faced significant issues, with about one in four failing catastrophically within their first year. To address this problem, satellite manufacturers established strict guidelines, including constructing satellites in clean environments and conducting various rigorous tests. These include:

- Controlled environments: Satellites were built in clean rooms to prevent contamination.

- Quality control programs: Rigorous programs ensure each component meets high standards.

- Piece-part screening: Individual parts were tested before assembly.

- Dynamic environmental testing: Simulates launch conditions to ensure durability.

In these efforts, the rate of early failures improved little. As a satellite's way of communicating its health became essential, it was realized that it is not always possible to compare data from factory tests directly to what happened in space.

6.1.2.1.
Advanced Testing Facilities

To address persistent reliability issues, satellite factories have evolved significantly, incorporating cutting-edge testing facilities. These integrated factories, some spanning over 600,000 ft^2, offer unrivaled production and testing capacities. They conduct multiple spacecraft-level thermal vacuum tests, antenna testing, and thermal stress testing simultaneously. Such capabilities enable shorter production cycles and reduced program risks. Manufacturers used massive thermal vacuum chambers to simulate the harsh conditions of space. These chambers, which can exceed 63,000 ft^3, allow for the simultaneous testing of multiple satellites. The horizontal orientation of these chambers facilitates swift satellite movement, minimizing vibrations and shocks.

A critical aspect of ensuring satellite reliability is prognostic analysis, which employs various algorithms to maintain data integrity under severe operating conditions. This analysis involves examining telemetry data for signs of accelerated aging, nonrepeatable transient events, and changes in normal equipment behavior. Prognostic analysis provided several benefits:

- Early failure detection: Identified potential issues before they lead to failures.
- Extended operational life: Helped in planning maintenance and repairs.

- Reduced launch delays: Ensured components are reliable before launch.
- Improved satellite service availability: Minimized downtime due to failures.

Historically, the aerospace industry followed a "Markov-based" reliability paradigm, treating failures as instantaneous and random. The Markov property implies that the future behavior of a system, given its current state, depends only on that current state and not on its past states. This approach was suitable for simple systems but proved inadequate for complex satellite systems. Understanding and predicting failures, especially in systems like the GPS, is crucial. During the early phases of GPS development, predicting the performance and potential failures of components, particularly atomic clocks, was a significant challenge. State-transition diagrams were used to represent the system's discrete states and their transitions.

The development of failure models and predictive algorithms shifted the industry toward a non-Markov reliability paradigm. This new paradigm recognizes accelerated aging and allows for proactive identification of equipment prone to premature failure. Unlike traditional reliability engineering, health management now considers operational data, identifying nonrepeatable transient events and predicting failures more accurately. Failure models based on premature failures inform model-based prognostic algorithms, enabling near-perfect reliability. Prognostic analysis, with its active reasoning and predictive algorithms, offers a promising approach to minimize premature failures and enhance overall equipment reliability. By adopting these advanced techniques, the aerospace industry can ensure the production of more reliable satellites.

6.2.
The Typical Detection Process

To develop a robust understanding of spacecraft telemetry, it is essential to recognize the various recurring patterns the spacecraft experiences. When telemetry data are normal, the differences between predicted and actual data are minor. Conversely, abnormal telemetry data result in much larger discrepancies. This observation forms the basis of prediction-based anomaly detection, which involves two main steps: predicting data and detecting abnormalities. We first predict the telemetry data and then compare them with the actual data. By subtracting the predicted data from the real telemetry data, we obtain the residuals. These residuals are then analyzed using a detection strategy to identify potential anomalies. Given that telemetry data often exhibit trends and periodic patterns, anomaly detection methods must account for both macro-level changes (overall trends) and micro-level specifics (individual time points).

6.2.1.
Macro-Level and Micro-Level Analysis

"Macro-level" and "Micro-level" refer to the different scales of analysis required for effective anomaly detection:

- Macro-level analysis: This broader analysis examines telemetry data to understand overall trends, patterns, and changes over extended periods.
- Micro-level analysis: This detailed analysis focuses on individual time points or specific events, aiming to detect detailed characteristics or anomalies at a more granular level.

For enhanced detection models, it is crucial to analyze the influence of specific time points on the data. Analyzing the entire dataset provides a big-picture view but may miss minor details. Conversely, using a sliding time window to create a local detection method captures these details, but may overlook broader trends. To address this, a combined strategy that considers different time scales is necessary. This approach ensures a comprehensive understanding of both the macro and micro aspects of telemetry data, improving the effectiveness of anomaly detection.

Telemetry data are often collected in a chronological sequence, capturing both temporal and spatial characteristics. Effectively modeling such data requires robust time series models that can handle these aspects. The structure of these methods typically follows a framework that incorporates both macro- and micro-level analysis. The entire process can be summarized as follows:

1. Data preparation: Telemetry data are divided into two parts. The first part, free of anomalies, serves as training data to train the model. The second part, testing data, may contain anomalies.

2. Model training: This involves training an ML model using the prepared training data.

3. Prediction: We employ testing data and the trained model to generate predicted data.

4. Residual data construction: Point-by-point subtraction of testing data and prediction data forms the residual data.

5. Anomaly detection: A strategy applies to detect anomalies within the residual to identify any abnormality.

6. Visualization: Presents anomaly detection results for clear interpretation.

When combined with command data, the spacecraft links the telemetry data it produces to the manual commands it receives, creating a multivariate time series. When predicting one-dimensional telemetry data, it is helpful to consider the command at each moment for improved accuracy. Let us consider a specific channel's telemetry data, denoted as X. At a given time, t, we can represent a set of m-dimensional vectors, $x(t) = \{x_1(t), x_2(t), \ldots, x_{m-1}(t), x_m(t)\}$.

In this representation, $x_m(t)$ signifies the telemetry data at the current time, while the others represent remote control instructions. We can also transform the telemetry dataset X for this channel into a tensor for its application in some DL algorithms (Figure 6.1). A tensor as a data structure will have three dimensions to represent the information:

- Time steps: Each time series data point is associated with a specific time step. For example, if you are monitoring temperature every hour, each hour's temperature reading would be a different time step.

- Samples: This dimension represents the various entities or instances in your dataset. Each entity can represent a unique time series. For instance, if you are collecting temperature data for different cities, each city's temperature data would be a separate sample.

- Features: The features dimension includes the variables or attributes measured at each time step. In time series data, this could be the different variables you are monitoring (e.g., temperature, humidity, and pressure).

Figure 6.1 Setting up telemetry data as a multivariate time series.

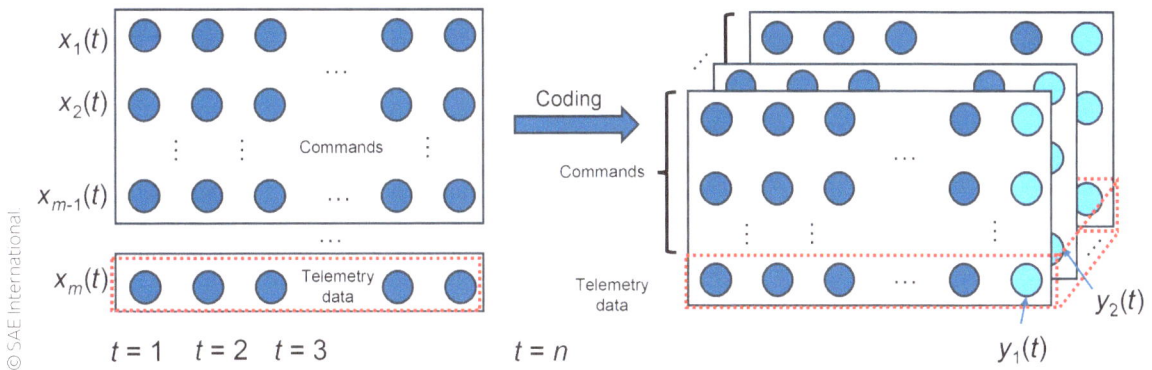

A data-driven approach can train individual models for each channel [6.1, 6.2]. We can then generate a predicted value $y(t)$ by inputting testing data and validating the result.

This difference, between the predicted value and the actual data, is denoted as $e(t)$ and is called residual:

$$e(t) = |y(t) - \hat{y}(t)|$$

The residual will have some noise because of errors in the individual points predicted by the trained model. To mitigate this, we often smooth it out either by averaging or using filters. But probably the most commonly used method is calculating the exponentially weighted moving average (EWMA), $e_s(t)$:

$$\alpha = \frac{2}{1 + span}, \quad span \geq 1$$

$$e_s(t) = \frac{e(t) + (1 - \alpha)e(t - 1) + \cdots + (1 - \alpha)^t e(0)}{1 + (1 - \alpha) + (1 - \alpha)^2 + \cdots + (1 - \alpha)^t}$$

The span shows the decay and depends on the length and characteristics of the testing data; usually, a value between 8 and 12 is suitable for analysis.

Earlier, we mentioned that detecting anomalies in the residual data usually involves setting a threshold, beyond which we identify points as anomalies. Directly specifying this threshold can be challenging. Therefore, it is worth designing a dynamic threshold to detect abnormal patterns using a sliding window. Each window will be linked to the mean (μ) and standard deviation (σ):

$$\varepsilon = \mu + z\sigma$$

In this approach, the value of z is determined by how the data are distributed within the window, and we select a minimum value based on experiments. Using a fixed-width window often proves effective for detecting anomalies because, when anomalies exist, the residual data exhibit significant local fluctuations. However, such an approach may miss out on slowly changing anomalies. To address this, we can introduce global detection as a supplement to assess the entire residual data (or with a larger window) using the adaptive threshold method mentioned earlier. Such a combination of anomaly detection methods at different time scales forms a comprehensive multiscale anomaly detection strategy.

Table 6.2 showcases various methods used for telemetry data analysis, such as Bayesian methods, dynamic Bayesian networks (DBNs), PCA, and DL with LSTM networks. These play a pivotal role in enhancing the reliability and safety of space missions. Bayesian methods, with their probabilistic approach, leverage prior knowledge and dynamically update probabilities, offering a robust framework for anomaly detection amid the inherent uncertainties of space operations.

Table 6.2 Common approaches for telemetry data analysis.

Method	Application suitability	Strengths	Considerations
Bayesian methods	Versatile for various domains, including spacecraft anomaly detection.	Good at handling uncertainty and incorporating prior knowledge.	Depends on accurate prior probabilities and understanding of system dynamics.
DBNs	Suitable for modeling complex systems with temporal dependencies.	Represents probabilistic relationships over time; adapts to system changes.	Modeling complexity; requires accurate dynamic models.
Principal component analysis (PCA)	Useful for dimensionality reduction in multivariate telemetry data.	Captures major variations; reduces dimensionality.	It may miss anomalies not represented in principal components.
DL— long short-term memory (LSTM)	Suitable for sequential data; captures temporal dependencies in telemetry time series.	Effective for long-term dependencies and patterns.	It needs sufficient training data; model complexity requires tuning.

Similarly, DBNs extend these capabilities to model temporal dependencies, crucial for understanding the evolving dynamics of spacecraft systems. PCA contributes by simplifying the analysis of multivariate telemetry data through dimensionality reduction, highlighting dominant patterns that show potential system issues. Meanwhile, LSTMs excel in capturing complex temporal patterns in sequential data, a key advantage for predicting system behaviors and potential failures.

These methodologies collectively represent significant advancements for health management systems, enabling more accurate predictions, diagnoses, and preventive measures that ensure mission success and spacecraft integrity. Bayesian methods and DBNs provide a solid foundation for incorporating and adapting to additional information, while PCA aids in managing the vast amounts of data generated by spacecraft systems. LSTMs, with their DL capabilities, offer further insights into temporal dependencies, although they require careful tuning and sufficient training data. These methods collectively enhance the capability of spacecraft health management systems to predict and diagnose potential issues accurately,

ensuring the longevity and success of space missions. Implementing these methods has led to significant advancements in autonomous spacecraft operations, although each method requires careful management of its own set of challenges.

6.2.2.
Modern Approaches
Integrating advanced analytical methods into spacecraft health management systems has significantly enhanced the capability to monitor, diagnose, and predict system health and anomalies [6.3–6.5]. Techniques such as hybrid feature selection for network intrusion detection, advanced concepts for intelligent vision systems, and survey on heterogeneous network traffic analysis have been applied in spacecraft telemetry analysis. These techniques were originally developed for network security, computer vision, and network traffic analysis applications. For instance, researchers have adapted the hybrid feature selection method to select relevant features in telemetry data, thereby enhancing anomaly detection accuracy while reducing the computational load and training time. This approach is influential in spacecraft health management, where the accuracy of

anomaly detection can be the difference between mission success and failure.

Similarly, the application of concepts from intelligent vision systems introduces advanced image processing algorithms to spacecraft telemetry, enabling a more sophisticated interpretation of visual data gained from spacecraft sensors. This is crucial for missions involving detailed planetary surface analysis or docking maneuvers where visual data accuracy directly influences operational decisions. The survey on heterogeneous network traffic analysis, although initially focused on network traffic, provides a framework for applying supervised and unsupervised learning models to spacecraft telemetry data. This approach offers a comprehensive method for managing the extensive and diverse telemetry traffic, ensuring effective monitoring and analysis. These adaptable methodologies significantly contribute to the evolution of health management systems, enhancing diagnostic capabilities, improving predictive accuracy, and streamlining data processing. This ensures higher mission safety and success rates (Table 6.3).

Table 6.3 Modern approaches used for telemetry data analysis.

Method	Application suitability	Strengths	Considerations
Hybrid feature selection for network intrusion detection	Applicable to feature selection in telemetry data for anomaly detection.	Enhances detection accuracy and reduces training time by selecting relevant features.	Focused on network security, but principles are adaptable to spacecraft telemetry.
Advanced concepts for intelligent vision systems	Useful for processing and interpreting complex telemetry data, especially involving visual data acquisition.	Offers advanced image processing algorithms for enhanced interpretation of visual telemetry data.	Primarily focused on computer vision and image processing outside the direct scope of telemetry.
Survey on heterogeneous network traffic analysis	Suitable for telemetry data analysis using supervised and unsupervised learning models.	Provides comprehensive insights into managing telemetry traffic effectively.	Originally intended for network traffic analysis, requiring adaptation for spacecraft telemetry.
New advances in intelligent decision technologies	Offers sophisticated tools for anomaly detection and decision support in systems.	Enhances the system with advanced models for telemetry analysis.	Focuses on intelligent decision technologies, necessitating specific adaptations for telemetry analysis.
ML-based anomaly detection	Suitable for satellite telemetry analysis.	Utilizes recurrent neural network (RNN), Moving Average (MA), and Fourier transform for feature extraction and AdaBoost for classification.	Requires comprehensive feature extraction and accurate anomaly classification.
Automatic anomaly detection techniques	Complements traditional OOL checking and monthly monitoring in spacecraft health monitoring.	Developed to detect anomalies missed by existing systems, showing promising results.	Operational assessment is needed for routine surveillance.
Multivariate latent techniques for anomaly detection and diagnosis	Focuses on characterizing normal system behavior for spacecraft systems.	Knowledge for monitoring acquired from telemetry data, using partial least squares discriminant analysis (PLSDA) for dimensionality reduction.	Requires robust information about the system's condition for effective monitoring.
Deep clustering-based local outlier probabilities (DCLOP)	It aims at detecting anomalies and extracting patterns from normal operational telemetry data.	Enhances the validity and accuracy of anomaly detection systems.	Assessed using actual cube satellite telemetry data, proving competitive effectiveness.

The methods highlight the importance of integrating ML and DL techniques to enhance the safety and reliability of space missions. They contribute to early warning systems for space missions by accurately detecting anomalies, ensuring prompt identification and resolution of potential failures.

6.3.
A Framework for Automatic Anomaly Detection

To address the various challenges mentioned in this chapter, major space agencies like ESA, NASA, Centre National d'Etudes Spatiales (CNES), Deutsches Zentrum für Luft- und Raumfahrt (DLR), Japan Aerospace Exploration Agency (JAXA), and companies like Airbus have been developing advanced automatic anomaly detection systems [6.6]. These efforts are also part of the ESA's Artificial Intelligence for Automation (A2I) Roadmap. More recently, the ESA proposed a Benchmark for Anomaly Detection in Satellite Telemetry (ESA-ADB), which aims to establish a new standard for ML-based satellite telemetry analysis and time series anomaly detection. It helps to identify the need for better evaluation approaches due to the scarcity of anomalies in flying missions and the flaws in existing public datasets and benchmarks. Different algorithms were tested, revealing the strengths and weaknesses of each, and the results highlight the challenges posed by ESA-ADB and the need for improved methods.

The ESA-ADB consisted of:

- ESA anomalies dataset (ESA-AD): A large-scale, structured dataset of real-life satellite telemetry from three ESA missions, manually

annotated and cross-verified by spacecraft operations engineers (SOEs) and ML experts.

- Evaluation pipeline: Designed for the practical needs of SOEs, introducing new metrics for satellite telemetry and simulating real operational scenarios.

- Benchmarking results: Results of algorithms tailored to meet space operations requirements.

Addressing these issues needed a robust, real-life dataset and a comprehensive evaluation pipeline. In this context, the ESA-AD:

- Comprises telemetry data from two main missions (Mission 1 and Mission 2), excluding a third due to insufficient anomalies. Includes 76 channels for Mission 1 and 100 for Mission 2, with specific channels monitored for anomalies.

- Contains more than 700 million data points per mission, offering a significantly larger volume than other public datasets.

6.3.1.
ESA-AD Evaluation Pipeline

The pipeline is designed to rigorously assess the performance of time series anomaly detection algorithms in a practical and comprehensive manner [6.7]. Here is a detailed breakdown of the evaluation pipeline:

- Datasets and data preparation:
  - ESA-AD: Collected from three ESA missions, containing real-life satellite telemetry data. It is manually annotated by SOEs and ML experts.
  - Training, validation, and test splits: Each mission is divided into halves, with the first half used for training and the second half

for testing. The last three months of the training set are used as a validation set.

- Resampling and standardization: The telemetry data are resampled to a uniform frequency using zero-order hold interpolation. Channels are standardized to zero mean and unit standard deviation based on nominal points in the training set.

- Hierarchical evaluation pipeline:

  - Hierarchical evaluation: Algorithms are compared one aspect at a time, from highest to lowest priority. This ensures a focus on the most critical aspects for SOEs and avoids the need to select weights for different aspects.

- Priority aspects and metrics:

  - Primary aspects: Focus on minimizing false alarms and correctly detecting anomalous events.

  - Corrected event-wise F0.5-score: Measures the balance between precision and recall, emphasizing precision to minimize false positives.

  - Subsystem-aware and channel-aware F-scores: Assess the identification of affected subsystems and channels.

- Secondary aspects: Address the practical utility of the algorithms.

  - Event-wise alarming precision: Evaluates the precision of detecting true events without redundant alarms.

  - Anomaly detection timing quality curve (ADTQC): Measures the accuracy of anomaly start time detection.

  - Modified affiliation-based F-score: Assesses the exact duration of anomalies and promotes proximity to ground truth.

- Benchmarking results (Table 6.4):

  - Algorithm selection: A mix of unsupervised and semisupervised algorithms are selected based on their ability to handle the specific challenges of satellite telemetry.

  - Implementation and adaptation: Selected algorithms are implemented or adapted within the TimeEval framework, ensuring compatibility and optimization for large datasets.

  - Comprehensive testing: Algorithms are tested on both lightweight subsets and full sets of channels. Lightweight subsets allow for initial experiments and familiarization, while full sets challenge the scalability and robustness of the algorithms.

Table 6.4 The benchmarking process.

Aspect	Why it is important?
Algorithm training	Algorithms are trained on the training set, which includes anomalies. This ensures that the algorithms can learn from both nominal and anomalous data.
Validation	The validation set is used to fine-tune hyperparameters and assess the initial performance of the algorithms.
Test set evaluation	Algorithms are evaluated on the test set, which contains previously unseen data. This step measures the generalization ability of the algorithms.
Metric calculation	The evaluation pipeline calculates the proposed metrics, focusing on primary aspects like minimizing false alarms and correctly detecting anomalies, as well as secondary aspects like detection timing and alarming precision.
Detailed reporting	The results of the benchmarking are reported in detail, including metric values for each algorithm and mission. Separate results for lightweight subsets and full sets of channels are provided.
Performance comparison	Algorithms are compared based on their scores across the different metrics, with a particular focus on their ability to handle real-life challenges of satellite telemetry.
Identification of strengths and weaknesses	The benchmarking process helps identify the strengths and weaknesses of each algorithm, providing insights into areas for improvement.
Open data science competitions	ESA-ADB plans to organize open data science competitions to stimulate the international community to develop better methods for satellite telemetry analysis.
Community contributions	Researchers and practitioners are encouraged to contribute new algorithms, propose improvements, and help extend the benchmark.

The ESA-ADB evaluation pipeline is designed to provide a comprehensive and realistic assessment of TSAD algorithms, addressing the specific challenges of satellite telemetry data. The benchmarking process ensures that algorithms are tested rigorously and objectively, promoting the development of robust and effective solutions for anomaly detection in satellite telemetry. The metrics proposed in the framework are tailored specifically to address the unique challenges of satellite telemetry anomaly detection, ensuring practical relevance and reliability:

- The corrected event-wise F0.5-score places a higher weight on precision (fewer false positives), which is crucial for practical applications in satellite operations where false alarms can be costly and disruptive. Also, having an event-level evaluation helps to estimate the performance at the event level rather than the sample level, making it more relevant for detecting distinct anomalies.

- The subsystem-aware and channel-aware F-scores assess the ability to identify affected subsystems and channels, which is critical for operators who need to pinpoint the source of anomalies accurately. By focusing on subsystems and channels separately, these metrics also prevent the loss of information that occurs when aggregating scores across channels.

- The event-wise alarming precision metric helps to evaluate the precision of detecting true events without redundant alarms, aligning with the need to avoid overwhelming operators with repeated notifications for the same anomaly.

- The ADTQC can measure the accuracy of anomaly start time detection, which is crucial for timely interventions. This metric's design reflects the preferences and operational needs of SOEs, accounting for the practical implications of early and late detections.

- The modified affiliation-based F-score helps to divide the ground truth into local zones, providing a detailed analysis of detection performance in different parts of the data. It also promotes detections that are close to the actual anomalies, enhancing practical utility by ensuring detections are relevant and timely.

These metrics offer general improvements over traditional metrics as they address the flaws of unrealistic anomaly density found in many traditional datasets, for example, they are designed to handle the specific complexities of satellite telemetry, such as high dimensionality, varying sampling rates, and noise. Setting strict and practical evaluation criteria, these metrics encourage the development of robust algorithms capable of performing well in real-world conditions [6.8]. A traditional sample-wise metric would often fail to account for the event-level significance of anomalies, leading to misleading performance evaluations. They can also be overoptimistic and may not be reflective of real-world operational needs. Another example is that metrics like areas under curves are not always suitable for the binary detection needs of SOEs, whereas the proposed metrics provide clear, actionable insights. As a result, the metrics proposed in the ESA-ADB framework seem to align more closely with the practical needs and operational realities of satellite anomaly detection, ensuring more reliable, relevant, and actionable evaluations.

6.3.2.
Interpretability

The ESA-ADB framework also addresses interpretability, which is becoming more and more relevant for practical applications in satellite telemetry. The framework offers a comprehensive framework for understanding and evaluating satellite telemetry anomalies through detailed annotations and a robust dataset structure. Manually annotated by SOEs and ML experts, ESA-AD provides a clear interpretability of what constitutes an anomaly, enhancing the detection results. The dataset's structure groups channels into subsystems and related channels, simplifying the identification of the impact and source of anomalies within the satellite's complex telemetry.

To further improve interpretability, the dataset employs subsystem-aware and channel-aware F-scores, which evaluate an algorithm's ability to identify affected subsystems and channels precisely. This approach avoids over-aggregation of scores across channels, offering detailed insights into specific parts of the system that are impacted. Additionally, the event-wise alarming precision metric reduces redundant alarms, maintaining clarity and making it easier for operators to respond to anomalies; for example, ADTQC can provide precise timing information, measuring the accuracy of anomaly start time detection. This timing information is crucial for understanding the context and potential causes of anomalies. Moreover, the modified affiliation-based F-score divides the ground truth into local zones and calculates precision and recall for each, promoting detections that are close to the actual anomalies and ensuring relevance and timeliness.

ESA-AD's transparent evaluation pipeline employs a hierarchical evaluation approach, focusing on the most critical aspects first. This structured method isolates and measures specific aspects of anomaly detection, offering clear and focused insights into algorithm performance.

The open-source nature of the dataset and its evaluation framework allows for reproducibility of results and continuous improvement through community contributions, ensuring that methods and metrics remain transparent and interpretable as they evolve. This comprehensive approach, supported by scholarly AI critique and insights, addresses the critical need for precise, interpretable anomaly detection in satellite telemetry. By focusing on detailed annotations, subsystem-specific evaluations, precise timing, and community-driven improvements, ESA-AD sets a high standard for the reliability and clarity of anomaly detection in space operations. This framework not only enhances our understanding of satellite health, but also provides a robust foundation for future advancements in the field. These enhancements would ensure the approach is not only effective in detecting anomalies but also practical and actionable in real-world aerospace applications.

from manufacturing tests to launch preparations. It is essential for monitoring and ensuring the functionality of equipment in the extreme conditions of space travel. Recent advancements have seen the application of data mining approaches to enhance the quick identification of system abnormalities, crucial for ground maintenance and ensuring reliable operations. Despite being critically important, the potential of telemetry in predicting equipment longevity is not fully utilized. Challenges persist in the industry due to cost concerns and the complexity of comprehensive telemetry systems. However, frameworks such as the ESA-ADB promise a significant leap in safety and reliability, predicting equipment failures with certainty and revolutionizing spacecraft maintenance paradigms. As the space industry continues to evolve, embracing these technological advancements in telemetry will be key to more secure, efficient, and reliable space missions.

6.4. Summary

Spacecraft operations in space depend on the reliability of their systems, often challenged by unexpected events like power failures or abnormal temperatures. Telemetry data play a key role in ensuring operational reliability, serving as an important link between spacecraft and ground operators by providing real-time insights into spacecraft conditions. These data are essential for both research and decision-making, aiding in the detection of anomalies that could show potential faults. Telemetry plays a crucial role throughout an asset's lifecycle,

References

6.1. Obied, M.A., Ghaleb, F.F.M., Hassanien, A.E., Abdelfattah, A.M.H. et al., "Deep Clustering-Based Anomaly Detection and Health Monitoring for Satellite Telemetry," *Big Data Cogn. Comput.* 7, no. 1 (2023): 39, doi:https://doi.org/10.3390/bdcc7010039.

6.2. Mohammad, B.R. and Hussein, W.M., "A Novel Approach of Health Monitoring and Anomaly Detection Applied to Spacecraft Telemetry Based on PLSDA Multivariate Latent Technique," in *15th International Workshop on Research and Education in Mechatronics (REM)*, El Gouna, Egypt, September 2014, 1-6, IEEE.

6.3. Manikandan, V.M., Karthikeyan, S., and Bhuvaneswari, T., "Hybrid Feature Selection for Network Intrusion Detection Using Data Mining," *International Journal of Scientific and Research Publications* 10 (2020): 929-936.

6.4. Blanc-Talon, J., Philips, W., Popescu, D., and Scheunders, P., *Advanced Concepts for Intelligent Vision Systems* (Berlin/Heidelberg: Springer, 2007).

6.5. Jayachitra, D. and Tamilselvi, J.J., "Survey on Heterogeneous Network Traffic Analysis with Supervised and Unsupervised Data Mining Techniques," *International Journal of Computer Science and Mobile Computing* 3, no. 7 (2014): 47-59.

6.6. Fuertes, S., Picart, G., Tourneret, J.Y., Chaari, L. et al., "Improving Spacecraft Health Monitoring with Automatic Anomaly Detection Techniques," in *14th International Conference on Space Operations*, Daejeon, South Korea, 2016, 2430.

6.7. Kotowski, K., Haskamp, C., Andrzejewski, J., Ruszczak, B. et al., "European Space Agency Benchmark for Anomaly Detection in Satellite Telemetry," arXiv preprint arXiv:2406.17826, 2024.

6.8. Cuéllar, S., Farías, G., Santos, M., and Alonso, F., "Preliminary Results on Anomaly Detection and Recognition in Spacecraft Telemetry," in *2022 IEEE International Conference on Automation/XXV Congress of the Chilean Association of Automatic Control (ICA-ACCA)*, Curicó, Chile, October 2022, 1-6, IEEE.

In engineering systems, faults often start as minor issues and can gradually become more severe over time. Small vibrations in a machine can become significant problems if ignored. To monitor these changes, it is essential to measure some markers during a system's operation. However, directly correlating these measurements with underlying problems in a physical model can be challenging [7.1].

Data-driven methods are valuable in this context, especially when there is a limited understanding of the system's inner workings. These methods analyze collected data to determine what is normal for the system, identify damage patterns, and assess the condition of individual components or the entire system. The choice of learning method depends on the data. If the data have labels, we can use supervised learning to detect patterns and make predictions. Unsupervised learning is useful for exploring the structure of unlabeled data and validating hypotheses. Unlike traditional methods that rely on physical models, data-driven approaches use mathematical models. They assign varying importance to different data points based on past information, making them effective for predicting imminent issues, such as when a component is close to failure. However, these methods require extensive, high-quality data, which can be challenging to get in fields like aerospace because of security concerns and the infrequency of component failures.

Data-driven methods often employ AI techniques, including neural networks, support vector machines (SVMs), clustering algorithms, and statistical methods like Gaussian process regression (GPR) and hidden Markov models (HMMs). They help to predict and understand system behaviors, especially in complex systems.

7.1.
AI vs. ML vs. DL

People frequently use and often interchangeably refer to the terms AI, ML, and DL. However, understanding their distinct roles and relationships is crucial:

- AI: AI represents a broad branch of computer science focused on creating systems capable of performing tasks that typically require human intelligence. This includes systems that emulate human intelligence based on fixed rules and databases, such as knowledge graphs and expert systems. AI encompasses ML systems that can adapt and change themselves when exposed to new data.

- ML: ML is a subset of AI. It is the study of algorithms and statistical models that enable computers to perform a task without explicit instructions or programming. ML systems learn and make decisions based on data, and they improve their performance over time through exposure to more data.

- DL: DL is a specialized subfield of ML, focusing on artificial neural networks (ANNs), particularly deep neural networks (DNNs). DNNs are neural networks with multiple layers, designed to mimic the biological neural networks found in animal brains. Despite being around for decades, the significance of neural networks has surged because of computational advancements. DNNs have achieved remarkable accuracy in various complex tasks, including image and sound recognition, recommender systems, and natural language processing.

Each of these fields, while interconnected, addresses different aspects and complexities of creating intelligent systems. AI is the broadest concept, encompassing any computer system that exhibits traits of human intelligence. The concept is intrinsically related to being data-driven. Developers, trainers, and deployers of AI systems build upon this core relationship. ML is a practical approach within AI, focusing on systems that learn from data. DL, as a subset of ML, delves into more complex architectures, like neural networks, to solve intricate problems. Understanding these distinctions is essential for accurately discussing and applying these technologies in various engineering fields [7.2].

We can typically categorize ML scenarios based on the nature of the task they aim to solve and the type of data they have for training and testing the algorithms. The three main categories include supervised learning, unsupervised learning, and reinforcement learning. Each category is associated with specific types of tasks:

- Supervised learning: The algorithm trains on a labeled dataset, where each input is paired with the correct output. The algorithm's goal is to learn and predict unseen data output.

- Classification: The aim is to categorize each input into discrete classes or labels. Examples include spam detection (classifying e-mails as spam or not) and image classification (identifying the content of images, such as cars, planes, or boats). Fault detection in systems like spacecraft can also be a classification task.

- Regression: This involves predicting a continuous output value based on input features. Common examples are predicting stock market prices or house prices based on attributes like the number of rooms or areas.

- Unsupervised learning: Algorithms train on unlabeled data, where the correct output for each input is not provided. This makes quantitative performance evaluation challenging. The goal is to discover patterns or structures in the data [7.3].

- Clustering: The aim is to group the dataset into homogeneous subsets or clusters.

- Dimensionality reduction: Techniques like PCA are used to reduce the dimensionality of high-dimensional data, simplifying the dataset while retaining essential information.

- Reinforcement learning: An agent learns to decide by performing actions in an

environment to achieve a goal. The agent receives feedback as rewards or penalties and learns to maximize cumulative rewards.

- Learning optimal strategies in games, robotic navigation and control, and optimizing decision-making processes in various applications.

Each of these learning paradigms has its unique methodologies, challenges, and applications, making them suitable for different problems in ML (Figure 7.1).

Figure 7.1 Overview of types of learning.

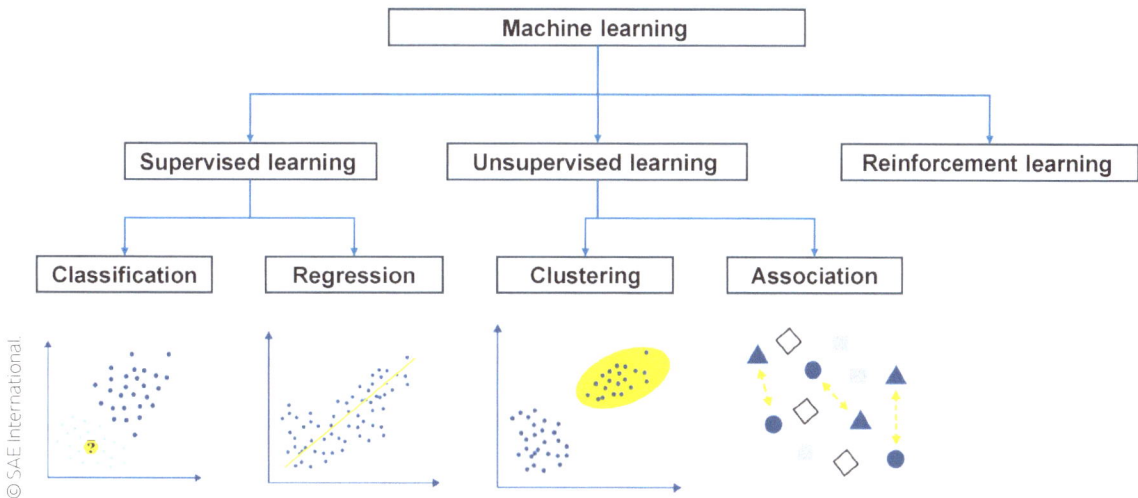

© SAE International

7.2.
Supervised Learning

7.2.1.
Linear and Polynomial Regression

Regression deals with predicting a continuous output based on input features. It is about establishing a relationship between variables and forecasting [7.4]. Linear regression is a statistical method widely used in supervised learning for predicting a continuous dependent variable based on one or more independent

variables. The core idea of linear regression is to establish a linear relationship between the input(s) and the output. This relationship is represented by a linear equation of the form $y = \beta_0 + \beta_1 x_1 + \beta_2 x_2 + \cdots + \beta_n x_n$, where y is the dependent variable, $x_1, x_2, \ldots, x_n$ are the independent variables, and $\beta_0, \beta_1, \ldots, \beta_n$ are the coefficients that the model learns during training. The coefficient β is known as the intercept, while $\beta_1, \ldots, \beta_n$ are the slopes that quantify the impact of each independent variable on the dependent variable.

In the training phase, linear regression involves finding the values of these coefficients that best fit the data. We can minimize the sum of the squared differences between the observed values and the values predicted by the linear model, which is known as ordinary least squares (OLS), to find the values of these coefficients that best fit the data. The resulting model can then predict the value of the dependent variable for new, unseen inputs. Many researchers prefer linear regression because of its simplicity and interpretability. It works best when the relationship between the variables is linear and the data is not too complex.

However, it can struggle with nonlinear relationships and overfitting when there are too many features. Despite these limitations, linear regression remains a fundamental and widely used technique in both statistics and ML (Figure 7.2).

Figure 7.2 (a) Linear regression and (b) polynomial regression working principle.

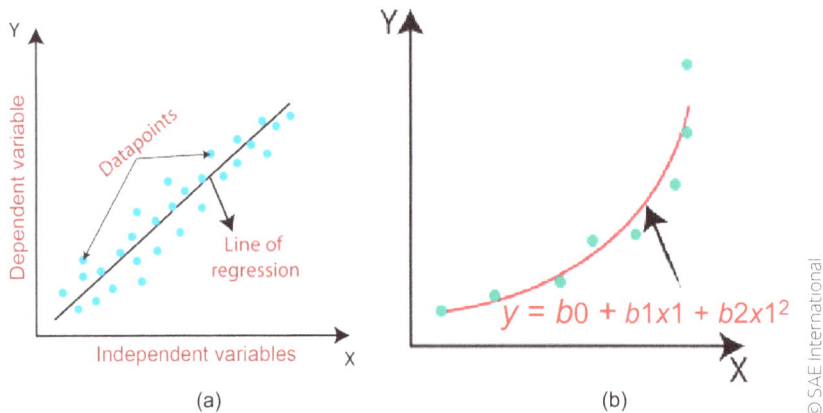

In polynomial regression, we model the relationship between the independent variable x and the dependent variable y as an nth-degree polynomial. Polynomial regression fits a nonlinear relationship between the value of x and the corresponding conditional mean of y, denoted $E(y|x)$. The model is expressed as: $y = \beta_0 + \beta_1 x + \beta_2 x_2 + \cdots + \beta_n x_n + \epsilon$.

Unlike linear regression, polynomial regression can model data where the relationship between the independent and dependent variables is curvilinear. Adjusting the polynomial degree increases model flexibility for diverse data patterns. However, higher-degree polynomials can often lead to overfitting, where the model captures noise in the data rather than the underlying trend. It is important to choose the degree of the polynomial carefully.

Where the data show a clear trend that is not linear, and a more complex model is needed to capture the relationship between variables accurately, polynomial regression proves to be useful. However, it is important to balance the model's complexity with the risk of overfitting.

7.2.2.
SVMs

SVMs are a type of supervised ML model highly effective for both classification and regression tasks. The core concept of SVMs is to identify the best decision boundaries, or hyperplanes, in a multidimensional space. These hyperplanes effectively separate different classes in the labeled training data based on their input features. Once established, SVMs can classify new data points into the correct categories, acting as a nonprobabilistic binary linear classifier.

SVMs come in two main types: hard margin and soft margin classifiers. Hard margin classifiers work well with linearly separable data, focusing on finding parallel hyperplanes that distinctly separate the classes. However, we use soft margin classifiers when data are not perfectly linearly separable. They aim to find the best hyperplanes for class separation while also using a balance parameter, λ, to allow some misclassification for improved overall performance. One of SVMs' key strengths is their strong generalization capabilities, often leading to superior performance compared to traditional ML methods, particularly in classification tasks. This makes them ideal for condition monitoring and fault diagnosis, where they excel in detecting patterns in signals and classifying them based on fault occurrence. SVMs are also efficient with fewer samples, maintaining accuracy even with limited fault samples.

The application of SVMs for monitoring the health of spacecraft components, diagnosing faults, and predicting the RUL of parts is widespread (Figure 7.3). Despite not being as commonly used as some other ML models, SVMs have exhibited robust performance across various domains. This robustness positions them as valuable tools for advanced diagnostic and prognostic applications in the aerospace industry, where accurate and reliable fault detection is crucial.

Figure 7.3 SVM working principle.

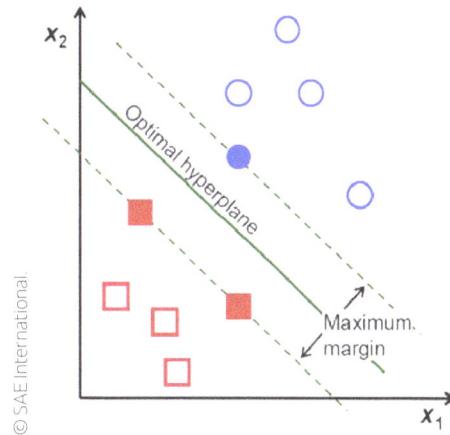

© SAE International.

7.2.3.
k-Nearest Neighbors (kNNs)

The kNN algorithm is a fundamental tool in ML, known for its simplicity and effectiveness, especially in classification tasks. Experts value the algorithm as a nonparametric learning algorithm that categorizes data based on similarity measures, helping to group similar data points from a training dataset. When classifying new data points, it considers proximity to a predefined number of the nearest neighbors, known as "k" points. The classification of a new data point is determined by a voting mechanism among these kNNs. The majority class among neighbors dictates the classification assigned to the new data point. Choosing the optimal k varies depending on the dataset and problem specifics. The right balance prevents overfitting (small k) and over-smoothing (large k). Also, kNN uses distance functions like Euclidean, Manhattan, and Minkowski to measure dissimilarity between data points.

An illustration of kNN's functionality shows how it classifies an unseen data point in linear and nonlinear datasets. For instance, a linear dataset with $k = 3$ can assign the new data point.

If two out of three nearest neighbors are in Class 1, it classifies as Class 1. The algorithm's versatility allows for wide usage across various industries, specifically for fault detection tasks. Additionally, it offers the possibility of combining it with other methods, like ANNs, to enhance accuracy in real-time fault prediction and offer a more robust solution.

7.2.4.
GPR

GPR is a sophisticated nonparametric Bayesian modeling technique used in ML. Unlike traditional methods that offer a single-point prediction, GPR provides a probability distribution, offering a more comprehensive view of potential outcomes. This technique is defined by two key components: a mean function and a covariance function. These functions collectively establish the relationships between the inputs and outputs of a system. The foundation of GPR lies in the Gaussian process, which is essentially a collection of random variables. Any finite number of these variables has a joint Gaussian distribution. This probabilistic approach makes GPR an effective tool for modeling and prediction, particularly in situations where uncertainty plays a significant role. One of the primary applications of GPR is in system monitoring, where it is used to predict future states and estimate the RUL of components. GPR's ability to provide a range of future states rather than a single prediction allows for a more nuanced understanding of a system's health and the uncertainties involved. This feature is especially valuable in prognostics, where the aim is to predict future degradation based on historical data.

However, GPR faces certain challenges. One significant challenge is its computational

intensity, which increases with the size of the training dataset. This can lead to memory-intensive operations and slow down the prediction process. To mitigate this, researchers have developed sparse approximation strategies that maintain GPR's efficiency while reducing computational demands.

Another challenge arises in handling time-series data in health monitoring. We need accurate surrogate models to approximate real-world phenomena. These models must be flexible enough to provide accurate approximations across various inputs and capable of extrapolating beyond existing data to predict future states. This task is complex, as the system's behavior during extrapolation may differ markedly from its behavior within the training data range. Addressing these challenges is crucial for leveraging GPR's full potential in predictive modeling and system health monitoring.

7.3.
Unsupervised Clustering

7.3.1.
Fuzzy C-Means (FCM)

FCM is a clustering technique that offers a more nuanced approach to organizing data compared with traditional clustering methods. Unlike conventional clustering, where each data point is assigned to a single cluster, FCM enables data points to belong to multiple clusters with varying degrees of membership. This flexibility proves particularly useful when the boundaries between clusters are not distinctly defined, as it allows for a more comprehensive view of the data.

The FCM algorithm operates through several key steps:

- Initialization: It begins by randomly selecting "c" cluster centers from the data points.

- Membership calculation: For each data point, FCM calculates its degree of membership in each cluster. This calculation is based on a formula that factors in the distance of the data point from the cluster center and a weighting exponent "m," which influences the level of cluster fuzziness.

- Cluster center update: The algorithm then updates the cluster centers. The degrees of membership of the data points influence this update, ensuring that the centers represent the "fuzzy" nature of the clusters.

- Cost function evaluation: FCM evaluates a cost function that incorporates the cluster centers, membership degrees, and distances between data points and cluster centers. The aim is to minimize this cost function, optimizing the cluster arrangement.

- Iteration: The algorithm iterates through steps 2 to 4, recalculating memberships and updating cluster centers until there is minimal change in the cost function or membership degrees, showing convergence.

In practical applications, FCM's ability to assign data points to multiple clusters is highly beneficial. For instance, in monitoring the health of machinery or systems, FCM can identify subtle patterns and trends that more rigid clustering methods might overlook. Early intervention or further investigation can more accurately assess a sensor reading that hovers between "normal" and "abnormal" states. This capability of FCM to handle ambiguity and provide a more detailed analysis makes it a valuable tool in various fields, especially where precision and nuanced understanding of data is crucial (Figure 7.4).

Figure 7.4 The center for each cluster is shown as a black X.

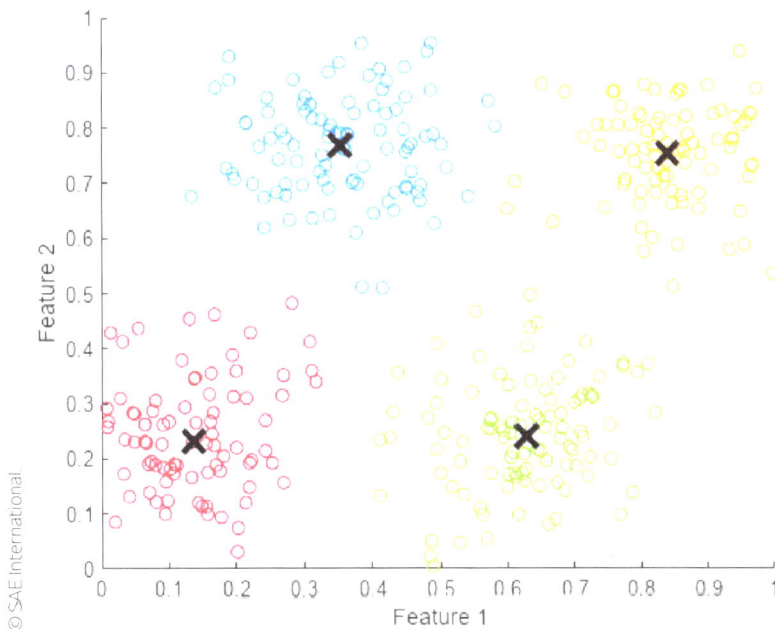

7.3.2.
K-means++

K-means is a common choice for its simplicity and efficiency. It partitions *n* observations into k clusters in which each observation belongs to the cluster with the nearest mean, serving as a prototype of the cluster. Objects within the same group are more similar to each other than to those in other groups. NASA's IMS uses clustering to analyze system data and identify these groups based on similarities in sensor readings and operational parameters. The process results in creating a model of normal system behavior for:

- Deviation detection: IMS continuously compares current system data against the established normal models. When data deviate significantly from these models, IMS identifies these deviations as anomalies, which may indicate potential system issues or failures.

- Sensitivity and specificity: The effectiveness of IMS in anomaly detection hinges on its ability to balance sensitivity (the ability to detect true anomalies) and specificity (the ability to ignore nonanomalous deviations). Fine-tuning these parameters is crucial for minimizing false alarms while ensuring no significant anomaly goes undetected.

K-means++ is an extension for choosing the initial values (or "seeds") for the K-means clustering algorithm. The standard K-means algorithm is sensitive to the initial choice of centroids, and a poor choice can cause suboptimal clustering. K-means++ improves the initialization phase, leading to better outcomes with the same computational complexity as the standard K-means algorithm.

Advantages of K-means++ are as follows:

- Improved initialization: By spreading out the initial centroids, K-means++ reduces the likelihood of converging to a local minimum that is not the global minimum.

- Better clustering results: Studies and practical applications have shown that K-means++ produces more accurate and consistent clustering results compared to random initialization.

- No additional computational complexity: Despite its improvements in the initialization phase, K-means++ does not significantly increase the computational complexity of the standard K-means algorithm. The most significant additional computation is in the initial selection of centroids, which is negligible compared to the entire clustering process.

7.4.
Probabilistic Models: HMMs

We use HMMs to represent systems that have underlying hidden states, which are not directly observable but can be inferred from observable sequences. These models extend Markov chains, a concept based on the principle that the future state of a system is determined solely by its current state, rather than the sequence of events leading up to it. The key elements of HMMs include:

- State transition probabilities (*a*): These probabilities define the likelihood of transitioning from one hidden state to another within the system.

- Observation probabilities (b): They describe the probability of observing a specific sequence when the system is in a particular hidden state.

- Initial state probabilities (π): These probabilities determine the likelihood of the system starting in each possible hidden state.

At each time step, the model transitions between hidden states based on the state transition probabilities. The model concurrently generates an observable sequence at each time step, influenced by the current hidden state and its associated observation probabilities. This process creates a series of observable events, each shaped by the underlying hidden states, which form the basis for further analysis and inference. To accomplish this, specialized algorithms, like the Viterbi algorithm, find the most probable path through the hidden states that result in the observed sequences. It is a key tool for decoding the hidden states in HMMs. Another essential algorithm is the forward–backward algorithm, which is used for calculating the posterior probabilities of the hidden states given the observed sequences. It helps to comprehend the likelihood of the system's hidden state at any point.

When modeling systems where the internal states are not directly observable but can be inferred from external observations, HMMs demonstrate powerful capabilities. Their ability to handle the probabilistic nature of state transitions and observations makes them suitable for a wide range of applications (Figure 7.5).

Figure 7.5 (a) A Markov chain; (b) HMM.

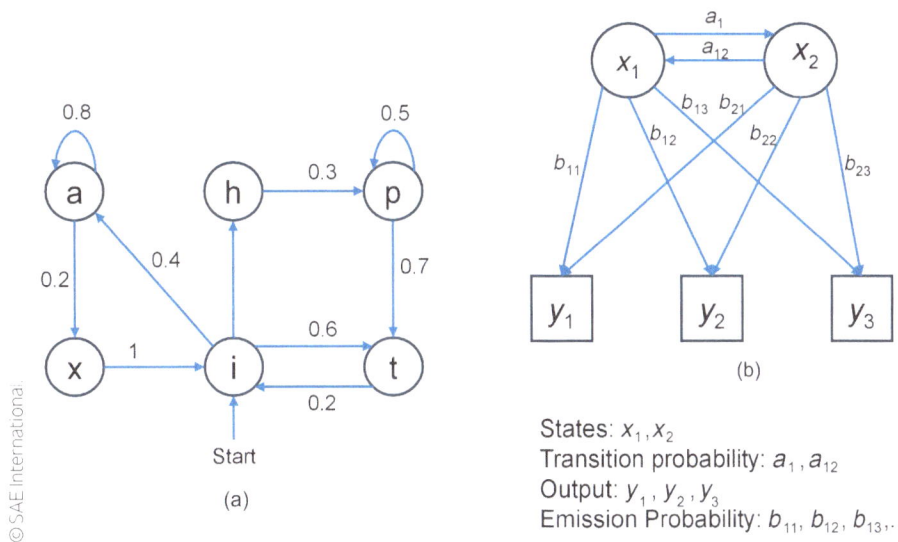

States: x_1, x_2
Transition probability: a_1, a_{12}
Output: y_1, y_2, y_3
Emission Probability: $b_{11}, b_{12}, b_{13},..$

© SAE International

7.5.
ANNs

ANNs are a type of ML algorithm inspired by the biological neural networks found in animal brains. They are particularly effective for pattern recognition tasks, making them valuable tools for diagnosing and predicting issues in various systems, including spacecraft. ANNs learn from data to produce specific outcomes, such as predicting the RUL of a spacecraft component or assessing its level of degradation. The fundamental components of an ANN are neurons, which are organized into layers. Each neuron in an ANN operates like this:

- Receiving inputs: A neuron receives multiple inputs, denoted as $p_1, p_2, p_3, ..., p_n$.

- Applying weights: Each input is multiplied by a corresponding weight. These weights are represented as $w_1, w_2, w_3, ..., w_n$. The weights are crucial as they are adjusted during the learning process to improve the accuracy of the ANN's output.

- Summation and bias addition: The weighted inputs are then summed together, and a bias term b is added. The bias helps the neuron adjust the activation function, which is crucial for learning.

- Activation function: It is used to process the sum of the weighted inputs and the bias. The activation function's role is to introduce nonlinearity into the output of a neuron. This is essential because it allows the network to learn and represent more complex patterns.

- Output generation: The output of the neuron, denoted as γ, results from applying the activation function.

One of the most commonly used activation functions in hidden layers of ANNs is the sigmoid function, denoted as $\sigma(\gamma)$. Mathematically, we represent the sigmoid function as:

$$\sigma(\gamma) = \frac{1}{1 + e^{-\gamma}}$$

This function maps the input γ (which can be any real value) to an output value between 0 and 1. This characteristic makes it particularly useful for tasks like binary classification.

Structure of a neuron: The structure of a typical ANN comprises three key components: an input layer, one or more hidden layers, and an output layer. These layers are composed of neurons interconnected by weights, which play a crucial role in the network's ability to process and learn from data:

- Input layer: This is the entry point of the network. The system takes in raw data, such as dataset features or image pixel values.

- Hidden layers: These layers perform the bulk of the computations through their neurons. Each neuron in these layers processes the inputs received from the previous layer, applies a weight to them, sums them up, adds a bias, and then passes them through an activation function. The number and size of layers can vary based on task complexity.

- Output layer: The final layer of the network produces the output. For a classification task, this might be the probabilities of different classes; for a regression task, it could be a continuous value.

The process of training an ANN involves adjusting the weights of the connections between neurons. We do this to minimize the error between the network's predictions and the actual outcomes. The most common method used for this purpose is backpropagation, coupled with an optimization algorithm like gradient descent. During backpropagation, the network calculates the error at the output and propagates it back, updating the weights to minimize this error.

DNNs are a specialized type of ANN characterized by having multiple hidden layers. This depth allows them to model complex relationships in the data. There are several types of DNNs, each suited for different tasks:

- Multilayer perceptrons (MLPs): These are the most basic form of DNNs, often used for simple classification and regression tasks.

- Convolutional neural networks (CNNs): They are designed for processing structured array data such as images, which makes them ideal for image recognition tasks.

- RNNs: They use sequential data like time series or natural language, as they have the unique feature of retaining information from previous inputs in the sequence.

Training these networks requires substantial data and computational resources. However, once trained, they are highly effective in tasks such as image recognition, natural language processing, and many others, demonstrating the power and versatility of neural networks in various fields of AI and ML.

RNNs are distinguished from other neural network architectures by their ability to process sequential data and maintain memory. RNNs achieve this memory by establishing connections between neurons that provide feedback, enabling them to effectively handle tasks involving time-series data or sequences by capturing temporal relationships within the data. The key features of RNNs include:

- Memory of previous outputs: RNNs have feedback connections that enable them to remember previous outputs. This is crucial for tasks requiring an understanding of temporal context.

- Elman network example: A classic RNN example is the Elman network, a simple three-layer network where the hidden layers receive inputs from both the current and previous time steps. The output at each time step depends on the current inputs, the previous outputs, associated weights, a bias vector, and an activation function.

- Short-term memory: This structure allows RNNs to maintain short-term memory, making them effective for tasks like language processing, where understanding the sequence of words is essential.

RNNs often face challenges in training because of these issues, where gradients become too small or too large, hindering the learning process. This is where we use LSTM networks, which incorporate gating mechanisms. These gates control the flow of information, allowing the network to keep important information over time and discard what's irrelevant. RNNs can also be bidirectional to enhance performance in sequence learning by processing data in both forward and backward directions, providing a more comprehensive understanding of the context.

The LSTM and bidirectional variants of RNNs have found extensive applications in areas such as fault prognosis and SHM, especially in the aerospace industry. Some notable applications include:

- Estimating RUL: RNNs can predict how long a component or system will function effectively, which is crucial for maintenance planning in aerospace and other industries.

- Predicting degradation of aircraft engines: By analyzing sequential data from engines, RNNs can predict their degradation over time, aiding in timely maintenance and repairs.

- Real-time fault diagnosis: RNNs can process data in real time to diagnose faults in systems, an essential capability for ensuring the safety and efficiency of aircraft and other complex machinery.

7.5.1.
Approximating Dynamics

Using ANNs to approximate the dynamics of spacecraft significantly enhances the accuracy and flexibility of onboard models, providing spacecraft with greater autonomy. There are various methods to tackle system identification and dynamics reconstruction using ANNs, each integrating the ANN model at different levels. Broadly categorizing these methods, we can classify them into:

- Fully neural dynamics learning: This method involves using an ANN to fully approximate the input–output structure of a spacecraft's dynamical model. The dynamics are fully captured by the weights and biases of the ANN.

- The ANN receives the current state and external output, and predicts the system state at the next time step or at the time derivative

of the state. This approach assumes that the ANN can approximate the dynamical function within a predefined error margin. Since RNNs can handle time-series data and maintain internal state memory, they are more beneficial than simpler feedforward networks.

- Dynamics acceleration reconstruction: This involves coupling the neural network with an estimation algorithm to reconstruct only the perturbation terms, using the linearized natural dynamics as a basis. RNNs, with their simple architecture and efficient evaluation time, are suitable for such onboard applications. They can efficiently handle the temporal evolution of states, making them ideal for dynamics identification.

- Parametric dynamics reconstruction: This method leverages the physics of the problem to get a representation that requires fitting only some parameters. Nonlinear autoregressive networks with exogenous inputs (NARX) models particularly suit dynamics reconstruction, as they can make predictions in a closed-loop architecture. NARX is a type of RNN that allows for nonlinear relationships by introducing a hidden layer of neurons between the input and output layers to capture complex patterns and relationships in the data.

Table 7.1 and the following discussion present a structured overview of how different AI network types can apply in spacecraft systems, emphasizing their specific architectures, training algorithms, and the challenges faced in space applications. Integrating AI in spacecraft systems offers promising advancements but also requires careful consideration of the unique challenges in space environments.

Table 7.1 Summary of typical approaches applied in spacecraft systems.

Approach	Architecture	Training algorithm	Space application scenarios	AI challenges
HMM	Hidden states, observable sequences	Viterbi algorithm, forward–backward algorithm	Sequence prediction, system behavior analysis	Inference complexity, handling hidden states
SVM	Hyperplanes in high-dimensional space	Sequential minimal optimization	Fault detection, system health monitoring	Computational intensity, resource requirements
GPR	Nonparametric, Bayesian approach	Optimization of covariance function	Estimating RUL, monitoring system health	Computational intensity, sparse approximation strategies
ANN	Multilayered, neurons organized in layers	Backpropagation, gradient descent	Proximity operations, relative pose estimation	Data limitations, generalization
RNN	Sequential data processing, feedback loops	Backpropagation through time	Time-series analysis, predictive maintenance	Handling time-series data, surrogate model accuracy

© SAE International

7.6.
Surrogate Models for Spacecraft Systems

Traditional linear designs, while effective in the past, face limitations when applied to the nonlinear and unpredictable environments of space missions. The adoption of ML, particularly nonlinear adaptive systems, offers enhanced capabilities and efficiency for these complex missions. From the discussion presented in this chapter, we identify three challenges:

- Nonlinear systems in space: Spacecraft operate in inherently nonlinear environments, characterized by varying gravitational fields, thermal extremes, and unpredictable celestial phenomena. Traditional linear systems may struggle to adapt to these complexities. Incorporating nonlinear, AI-based systems in spacecraft can significantly improve their ability to navigate and operate in space. These systems can better handle the dynamic and complex nature of space environments, leading to more efficient and successful missions.

- Autonomous systems: AI-driven autonomous systems can make real-time decisions based on the spacecraft's health data, detecting anomalies and potential failures early. The AI can predict potential system failures or component degradations, allowing for preemptive maintenance actions, thus extending the lifespan of spacecraft components.

- Online learning: Spacecraft can continuously update their operational models based on real-time environmental data through online learning, resulting in improved accuracy in predictions and decision-making. The continuous learning from new data allows for refining the mathematical models that govern spacecraft operations, ensuring they remain accurate and relevant.

The obvious challenge here is associated with the limited data available for training AI models, often requiring algorithms that can learn effectively from small datasets. Computational constraints that restrict onboard capabilities make this worse. Failures of these methods can have critical consequences, raising doubts about their robustness and reliability.

7.6.1.
A Dynamic Model for a Spacecraft

Modeling the dynamics of a spacecraft is a critical aspect of spacecraft design and operation. This process involves predicting how various factors, such as external environmental conditions and internal heat generation, affect the spacecraft's temperature. Extending this modeling to encompass the entire dynamical model of the spacecraft further enhances the understanding and control of the spacecraft's behavior in its operational environment.

For example, an ANN was used to model the thermal dynamics of a spacecraft (Figure 7.6). The model can be trained using data that comprises different spacecraft states and their corresponding thermal readings. The ANN learned to predict the thermal response of the spacecraft under different operational conditions, such as varying external temperatures, internal heat generation, and changes in spacecraft orientation. Such models can also predict potential thermal issues, like overheating of components, well in advance, allowing for preemptive measures. The model's ability to adapt to new data ensures it remains accurate over the spacecraft's lifetime and is based on the requirement that components must operate within their temperature limits, preventing overheating or underheating. The goal is to maintain optimal temperature ranges for all components, considering factors such as solar radiation, internal heat sources, and the thermal properties of spacecraft materials.

Figure 7.6 An ANN.

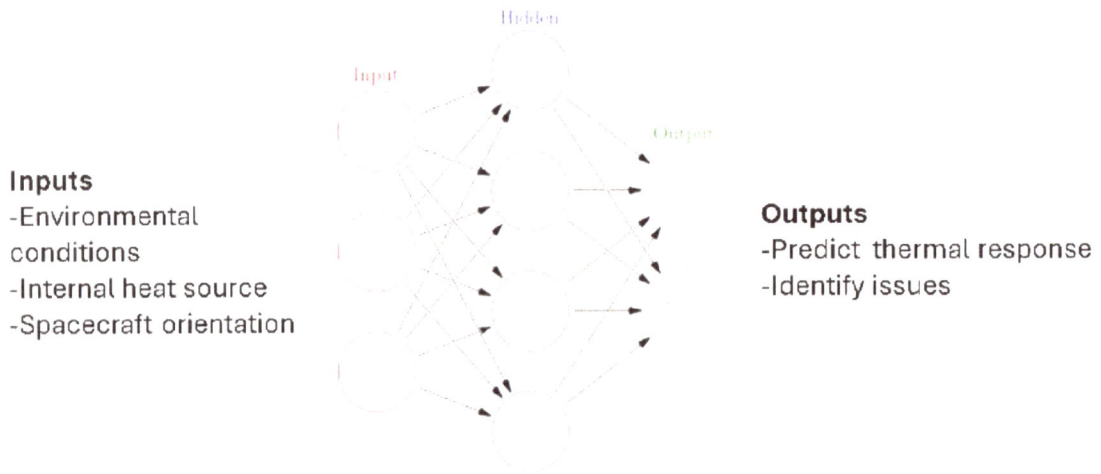

Inputs
-Environmental conditions
-Internal heat source
-Spacecraft orientation

Outputs
-Predict thermal response
-Identify issues

© SAE International.

In operation, the model receives real-time data about the spacecraft's state and predicts its thermal response, allowing for proactive measures to be taken if potential thermal issues are detected, such as adjusting the spacecraft's orientation to optimize thermal regulation or modifying the operation of heat-generating components. The model's adaptability to new data ensures its predictions remain accurate over the spacecraft's lifetime. This is crucial, as spacecraft components must operate within specific temperature limits to prevent damage.

The following are the challenges:

- One of the key challenges is accurately modeling the complex interactions between various heat sources and sinks.
- Another challenge is adjusting the model to time changes, like thermal insulation degradation.

By building on the initial model, we can extend the approach to other spacecraft systems. For instance, we can model the dynamic interactions between the thermal system and propulsion or power systems. A paper by Kumar et al. [7.5] demonstrated the use of RNNs to model the entire dynamical behavior, including its orbital mechanics, attitude control, and environmental interactions. This comprehensive model allows for simulations of various mission scenarios, enabling better planning and decision-making.

If the model accurately represents the spacecraft's behavior, it can autonomously make decisions in real time. For example, the model can autonomously adjust the spacecraft's orientation to optimize thermal regulation based on predicted solar exposure. One challenge is ensuring the model's accuracy with limited data. Hybrid modeling approaches can address this, combining physics-based models with AI techniques. Another challenge is the computational demand of such models, especially for onboard processing. Techniques such as model pruning and efficient neural network architectures can mitigate this issue.

7.6.2.
Approximating Unmodeled Dynamics

The second method for modeling spacecraft dynamics employs ANNs, specifically state estimators to learn and output an estimate of the disturbance/mismodeled terms, in combination with an adaptive filter. This approach is designed to enhance state and disturbance estimation by integrating classical techniques with modern AI-based methods. For example, a radial basis function neural network-aided adaptive extended Kalman filter (RBFNN-AEKF) approach can be particularly useful in scenarios where the system dynamics are nonlinear and subject to uncertainties or external disturbances. This would involve estimating and predicting the nonlinear states that ensure the functionality and longevity of onboard systems and instruments. Neural networks have also been employed to approximate complex dynamical systems, such as the chaotic three-body problem, enhancing spacecraft dynamics modeling [7.5].

The methodology comprises three key components:

- Capturing the nonlinear dynamics: A network can capture the nonlinear relationships within the system data, for example, CNN, LSTM, GPR, or radial basis function neural networks.
- Adaptive state estimation: The aim is to continuously update the state estimates of the model in an adaptive manner, incorporating the estimations from the previous step. This would allow for real-time adjustments, leading to more accurate predictions, for example, PF, moving horizon estimation, or EKF.
- Online learning and adaptation: An online learning ability can help a model adapt to new environments, such as entering different orbits with varying solar exposure or changes in onboard activity that affect heat generation. This adaptability ensures that the health management mechanism remains effective even as the system evolves or encounters new types of disturbances.

By accurately estimating system states and modeling nonlinear dynamics, such a setup can not only detect anomalies that deviate from normal operational patterns but can help isolate the source of the fault by analyzing which aspects of the system dynamics are contributing to the observed anomalies. Beyond immediate fault detection, this setup can also predict potential future failures by identifying trends or patterns that precede faults. Such predictive capability enables proactive maintenance action, reducing downtime and preventing catastrophic failures. Depending on the design, the adaptive state estimation component can effectively handle measurement and process noise. This robustness ensures that any corrupt data do not easily mislead the fault detection mechanism, leading to more accurate and reliable fault identification.

7.6.3.
Parametric Reconstruction

The third method focuses on parameter reconstruction using ANNs for refining uncertain parameters in dynamical models. This approach is important for adapting to environmental uncertainties that affect primary system constants, for example, inertia parameters, spherical harmonics, or drag coefficients. Two approaches exist for this problem:

- Using MLP neural networks: Aerospace engineers have increasingly applied the use of MLP neural networks, a class of DNNs, for tasks that require high levels of abstraction from input data. Chu et al.'s work exemplifies this application, where MLPs are used to estimate the inertia parameters of spacecraft— a critical factor in the control and navigation of spacecraft. The model inputs include angular rates and control torque, which are easily measurable in a spacecraft's operational

environment. Included in the output is the inertia tensor, which is a fundamental property that influences the spacecraft's attitude dynamics. DL's potential in capturing complex, nonlinear relationships inherent in spacecraft dynamics is underscored by the MLP model's ability to extract higher abstract features like the inertia tensor from raw data. This process not only aids in accurate spacecraft modeling, but also enhances the precision of simulation and control algorithms, contributing to more efficient mission planning and execution.

- Using RNNs and Hopfield neural networks (HNNs): RNNs and HNNs offer another sophisticated approach to modeling spacecraft dynamics, especially for parameter reconstruction. Unlike MLPs, RNNs have a temporal dynamic behavior because of their feedback connections, making them particularly suited for time-series data analysis, such as tracking the evolution of spacecraft states. HNNs, with their energy-minimizing nature, are adept at solving optimization and combinatorial problems, which are prevalent in spacecraft dynamics modeling. In the context of parameter reconstruction, researchers often reformulate the dynamical model of a spacecraft into a linear-in-parameters form. This reformulation transforms the identification of dynamical parameters into an optimization problem, where the aim is to minimize the discrepancy between the model predictions and actual observations. HNNs excel in this area by defining a weight matrix and a bias vector that encapsulate the relationship between the model inputs (e.g., spacecraft states and control inputs) and the desired outputs (e.g., estimated dynamical parameters). This capability allows for the efficient solving of combinatorial problems inherent in system identification tasks, offering a robust method for refining the models that govern spacecraft behavior.

By enabling more accurate estimation of critical parameters and facilitating the real-time adaptation of models to observed data, these ML techniques can substantially improve the predictive maintenance capabilities of spacecraft. This leads to enhanced mission safety, operational efficiency, and cost-effectiveness by allowing for early detection of potential issues and facilitating timely corrective actions. The ability of these models to learn from data and adapt to new environments makes them invaluable for long-duration space missions, where spacecraft may encounter unforeseen conditions. As ML techniques continue to evolve, experts expect to integrate them into spacecraft design and operation, yielding even greater benefits and pushing the boundaries of what is possible in space exploration and utilization.

7.7.
Summary

This chapter discussed data-driven reasoning for fault detection and prediction in mechanical systems, emphasizing the transition from minor issues to severe problems if not addressed timely. It highlights the challenge of correlating measurements to underlying issues because of the complexity of operations. Data-driven

methods prove to be especially useful when the system's inner workings are not fully understood. They allow for the analysis of collected data, identification of normal operation patterns, damage patterns, and assessment of component conditions. Applications of supervised and unsupervised learning based on data availability contrast traditional physical model reliance with mathematical models that learn from data to predict failures. These techniques, including neural networks and statistical methods, play a crucial role in these predictive models, offering insights into system behaviors and enhancing operational reliability in complex mechanical systems.

References

7.1. Lei, Y., Li, N., and Guo, L., "A Review on Empirical Mode Decomposition in Fault Diagnosis of Rotating Machinery," *Mechanical Systems and Signal Processing* 35, no. 1-2 (2018): 108-126, doi:https://doi.org/10.1016/j.ymssp.2012.09.015.

7.2. Bishop, C.M., *Pattern Recognition and Machine Learning* (New York: Springer, 2006).

7.3. Johnson, M.E. and Khoshgoftaar, T.M., "Anomaly Detection in Spacecraft Telemetry Data Using Isolation Forests," *IEEE Transactions on Aerospace and Electronic Systems* 56, no. 3 (2020): 1920-1934.

7.4. Lee, S. and Kim, J.H., "Fault Detection in Spacecraft Attitude Control Systems Using Supervised Learning," *Journal of Spacecraft and Rockets* 58, no. 2 (2021): 456-467.

7.5. Kumar, P., Das, A., and Gupta, D., "Differential Euler: Designing a Neural Network Approximator to Solve the Chaotic Three-Body Problem," arXiv preprint arXiv:2101.08486, 2021.

Probabilistic design has been a cornerstone of engineering for over 50 years, incorporating uncertainty into design analysis to help create optimal solutions. These methods assess the risks associated with space launch vehicle failure modes and determine the likelihood and consequences of failure scenarios in complex systems, such as NASA's Constellation Ares I launch vehicle. A broader set of tools in system safety, reliability, and risk assessment, including qualitative and quantitative techniques such as hazard analysis (HA), FMEA, FTA, and probabilistic risk assessment (PRA), provides a comprehensive view of system risk and interactions with other systems and the environment. Reliability engineering greatly benefits from such tools, especially when dealing with complex and high-risk systems like spacecraft.

Achieving increased reliability and operability within tighter budgets and shorter development schedules is necessary in the current economic and space launch climate. As part of the ongoing risk management of various space programs, we perform HA, reliability assessment, and PRA at the beginning of the design phase to influence design and operational decisions. This chapter examines how these approaches contribute to predicting and improving the dependability of spacecraft components and systems. Identifying potential failures and their probabilities is crucial for designing spacecraft that are robust and reliable.

8.1.
Importance of Probabilistic Design

Spacecraft systems deal with complex systems that must perform reliably under a wide range of conditions. One of the key challenges is to design systems that can withstand uncertainties and unexpected events. This is where probabilistic design comes into play. This is important because probabilistic design is used for the following sections.

8.1.1.
Handling Uncertainty

Traditional engineering design methods typically rely on deterministic approaches, assuming fixed values for various parameters involved in the design process. These parameters might include material properties (such as strength, elasticity, and thermal conductivity), environmental conditions (such as temperature, humidity, and radiation levels), and operational stresses (such as loads, pressures, and vibration levels). For instance, a traditional design might assume that a material's tensile strength is exactly 500 MPa or that the ambient temperature will always be 25°C. While these fixed values simplify the design process, they fail to account for the variability and uncertainty inherent in real-world conditions.

There are several drawbacks to relying solely on traditional design methods. First, to ensure safety, traditional methods often add large safety margins. This can lead to overly conservative designs that are heavier, more expensive, and less efficient than necessary. Second, if actual conditions deviate from the assumed values, the design might fail to perform as expected, potentially leading to catastrophic failures. Lastly,

fixed-parameter designs lack the flexibility to adapt to changing conditions or unexpected events, which can be critical in dynamic environments like space.

Probabilistic design, on the other hand, incorporates the variability and uncertainty of parameters into the design process. Instead of assuming fixed values, probabilistic design treats each parameter as a range of possible values described by probability distributions. This approach offers a more realistic and comprehensive analysis of the system's behavior under various conditions.

8.1.1.1.
Key Elements of Probabilistic Design

1. Probability distributions: Parameters are modeled using probability distributions that reflect their inherent variability. For example, a normal distribution might describe the tensile strength of a material with a mean of 500 MPa and a standard deviation of 50 MPa.

2. Monte Carlo simulations: Monte Carlo simulations are often used to perform probabilistic analysis. This technique involves running thousands or even millions of simulations, each time sampling values from the probability distributions of the input parameters. The results provide a statistical distribution of the outcomes.

3. Risk and reliability analysis [8.1]: Probabilistic design enables the calculation of risk metrics, such as the probability of failure or the reliability of the system over time. These metrics help engineers understand the likelihood and consequences of different failure scenarios.

4. Sensitivity analysis: Sensitivity analysis identifies which parameters have the most significant impact on the system's

performance. This information helps prioritize areas for improvement and risk mitigation.

The above list offers several advantages. By accounting for the variability in material properties, environmental conditions, and operational stresses, it provides a more accurate representation of the real-world performance of the system [8.2]. This leads to optimized designs that achieve a balance between performance, safety, and cost. Probabilistic methods allow for the identification of optimal parameter combinations that minimize the probability of failure while maximizing efficiency. Understanding the probability and consequences of different failure modes allows for better risk management and decision-making, enabling engineers to develop targeted strategies to mitigate the most critical risks [8.3]. Probabilistic design produces systems that are more adaptable to changing conditions and unexpected events, which is particularly important in the harsh and unpredictable environment of space.

Consider the structural design of a spacecraft. Traditional design might assume a fixed load and design the structure to withstand that load with a safety margin. Probabilistic design, however, would consider the variability in loads because of different mission profiles, material properties that might degrade over time, and environmental factors like micrometeoroid impacts. For example, instead of a single value for the tensile strength, a probabilistic approach might use a normal distribution based on material testing data. Thermal stresses might be modeled using a distribution that reflects the temperature variations the spacecraft will experience in orbit. A distribution based on historical launch data and simulations might represent launch loads. By running Monte Carlo simulations with these

distributions, engineers can get a distribution of the stress responses of the structure. They can then calculate the probability of structural failure under various scenarios and make informed decisions about where to reinforce the structure or which materials to select.

8.1.1.2.
Risk Mitigation

One of the key benefits of probabilistic design is its ability to consider the probability of different failure scenarios. This allows engineers to identify potential risks early in the design process. Traditional design methods might only highlight issues after a failure has occurred or under specific conditions, but probabilistic design provides a comprehensive view of possible failure modes and their likelihoods. By analyzing a range of outcomes, engineers can detect vulnerabilities that might not be apparent with a deterministic approach. For example, engineers can identify and address certain materials that have a higher probability of failing under specific stress conditions before building or deploying the system [8.4]. This early detection is crucial because it allows for adjustments in the design phase, which is generally less costly and less time-consuming than making changes later in the development cycle.

Once engineers identify potential risks, they can implement measures to mitigate them. These measures can include design modifications, material substitutions, or the addition of redundant systems to enhance reliability [8.5]. For instance, if a probabilistic analysis reveals that a component is likely to fail under extreme temperatures, engineers might choose a different material with better thermal resistance or add insulation to protect the component. This proactive approach significantly enhances the safety and reliability of the system. By addressing potential failures before they occur, engineers can prevent accidents and reduce the risk of catastrophic failures. This is especially important in high-stakes industries like aerospace, where the consequences of failure can be severe.

Consider a spacecraft that will operate in a region with a high risk of micrometeoroid impact. A PRA might show a significant probability of hull penetration over the mission duration. With this knowledge, engineers can take several mitigation steps. They might design the spacecraft with thicker or more advanced shielding materials, add redundant layers of protection, or develop procedures for repairing damage in space. These actions, informed by the PRA, help ensure the spacecraft remains operational and safe throughout its mission.

8.1.1.3.
Optimization
Probabilistic methods play a crucial role in optimizing design parameters to achieve the best possible performance while minimizing the likelihood of failure [8.6]. Traditional design approaches often focus on meeting minimum requirements with added safety margins, which can cause conservative and suboptimal solutions [8.7]. In contrast, probabilistic design methods enable engineers to explore a wider range of design possibilities and identify the most efficient and effective solutions. One of the key advantages of probabilistic design is its ability to balance performance and reliability. By treating each parameter as a probability distribution, engineers can evaluate how different combinations of these parameters affect the overall performance and reliability of the system by treating each parameter as a probability

distribution. This approach allows for a more nuanced consideration of both the performance goals and the risk of failure simultaneously.

Probabilistic methods facilitate the exploration of a vast design space. For instance, in the design of an aerospace component, engineers might consider various materials, geometries, and manufacturing processes. Probabilistic simulations, such as Monte Carlo simulations, can evaluate thousands of potential designs, each with different combinations of these variables. The results provide insights into which designs offer the best trade-offs between performance metrics (like strength and weight) and reliability measures (like the probability of failure under different conditions). Optimization through probabilistic methods is an iterative process. Engineers start with initial design parameters and use probabilistic analysis to identify weaknesses and areas for improvement. They can then adjust the parameters and reevaluate the design, continually refining it to achieve the optimal balance of performance and reliability. This iterative approach ensures that the final design is robust and well-suited to the operating environment.

Consider the design of an aircraft wing. Traditional design might result in a wing that is heavier than necessary to ensure it can withstand all possible loads. Using probabilistic methods, engineers can model the variability in loads because of factors like turbulence, material inconsistencies, and operational stresses. By simulating these conditions, they can identify a design that uses lighter materials or a more efficient structure, which still meets all safety requirements but with improved performance in terms of fuel efficiency and maneuverability. In real-world applications, probabilistic optimization has led to significant improvements. For example, in the automotive industry, probabilistic methods have been used to design lighter and more fuel-efficient engines without compromising durability. In the aerospace industry, these methods have contributed to the development of more efficient and reliable spacecraft components, enhancing mission success rates and reducing costs.

8.1.1.4.
Informed Decision-Making

Probabilistic design methods provide engineers with a comprehensive understanding of the range of possible outcomes and their associated probabilities. Unlike traditional deterministic methods, which often present a single, fixed outcome, probabilistic approaches offer a spectrum of potential scenarios. This detailed insight into the likelihood of various events occurring enables engineers to anticipate and plan for different situations more effectively. With a clearer picture of the risks and probabilities, engineers can make better-informed decisions throughout the design and development process. For instance, they can evaluate whether a proposed design modification will significantly reduce the probability of failure or if it will have minimal impact. This level of understanding allows for more strategic and rational decision-making, focusing on resources and efforts where they will be most effective.

One of the significant benefits of probabilistic design is its ability to balance safety and cost-effectiveness. Engineers can use probabilistic methods to identify designs that meet stringent safety standards while also being economically viable. By quantifying the probabilities of different failure modes, engineers can assess the cost-benefit ratio of various design choices. This ensures that cost savings do not compromise safety, but it also prevents unnecessary expenditures on overly conservative designs. Probabilistic analysis helps prioritize which design adjustments will have the most substantial impact on overall system reliability and performance. For example, if the analysis shows that a particular component has a high probability of failure under specific conditions, we can allocate resources to redesign or reinforce that component. Conversely, if a component has a low probability of failure, it might not require immediate attention, allowing for more efficient use of time and budget.

Consider the development of a satellite. Probabilistic methods can model the satellite's performance and reliability under various operational conditions, such as exposure to radiation, mechanical stresses during launch, and thermal cycles in orbit. By understanding the range of potential outcomes and their probabilities, engineers can make informed decisions about materials, shielding, and redundancy systems. This leads to a satellite design that is both robust and cost-effective, maximizing mission success while staying within budget constraints. In the real world, informed decision-making through probabilistic design has led to significant advancements. For instance, in the pharmaceutical industry, probabilistic models are used to optimize drug development processes, balancing efficacy and safety with production costs. Similarly, in civil engineering, probabilistic methods help design resilient infrastructure that can withstand natural disasters without excessive costs.

The rest of the chapter will present the tools used for system safety, reliability, and risk assessment.

8.2.
Probabilistic Design Analysis (PDA)

PDA offers a realistic approach to engineering design by considering the uncertainty inherent in the analysis process. Unlike traditional methods that provide a single value, it treats each variable as a probability distribution with a range of values. This allows for the identification of parameter combinations that lead to higher failure probabilities, offering opportunities for design improvements to mitigate risks. The core elements of PDA include:

- Underlying analysis assumptions: The foundational assumptions guiding the analysis.

- Core equations: Mathematical equations describing the system's physical behavior.

- Failure criterion: Conditions under which failure is considered to occur.

- Input and output parameters: Variables and their ranges that feed into and result from the model.

PDA often involves Monte Carlo simulations to compute the probability of critical failures under assumed uncertainties. These models incorporate core equations to describe the system's physics and calculate failure probabilities with a uniform distribution for parameter uncertainty. The fidelity of a PDA model depends on understanding the problem, designing with maturity, and accurately modeling the involved physics. These models are typically first-order simplified analyses, differentiating them from detailed component and subsystem reliability models used in various disciplines. This provides a more realistic analysis by accounting for uncertainty and variability in design parameters. It allows for the identification of high-risk parameter combinations and enables designers to make adjustments to avoid failure. This probabilistic approach leads to more robust and reliable designs, enhancing the overall safety and performance of complex systems like spacecraft. However, there are some challenges that need to be addressed, including:

- Determining input parameters: Identifying which parameters vary and quantifying their variations.

- Defining the failure criterion: Establishing a failure criterion based on concepts like burden versus capability (or stress versus strength) to compute failure probabilities accurately.

8.2.1.
Case Study: System Integration Failure Analysis (SIFA) on Ares I Launch Vehicle

Launch vehicles play a key role in the cost-effective space transportation infrastructure of the space programs. Such systems enable human exploration missions to return to the moon and venture onward to Mars and other destinations in the solar system. NASA was developing, testing, and assessing HW and systems for the Ares I crew launch vehicle during the years leading up to the eventual cancellation in 2010 (Figure 8.1). These efforts were part of NASA's rigorous process to ensure the safety and reliability of the vehicle intended for crewed missions to space. The SIFA team used PDA to assess the reliability of these vehicles, involving feasibility studies, model development, data collection, sensitivity analysis, and detailed reporting. This process helped quantify failure probabilities and support the overall loss of crew (LOC) and loss of mission (LOM) assessments.

Figure 8.1 The Ares I Crew Launch Vehicle can carry four to six astronauts to Earth orbit and has a 25-ton payload capacity for delivering supplies to the ISS or parking payloads in orbit.

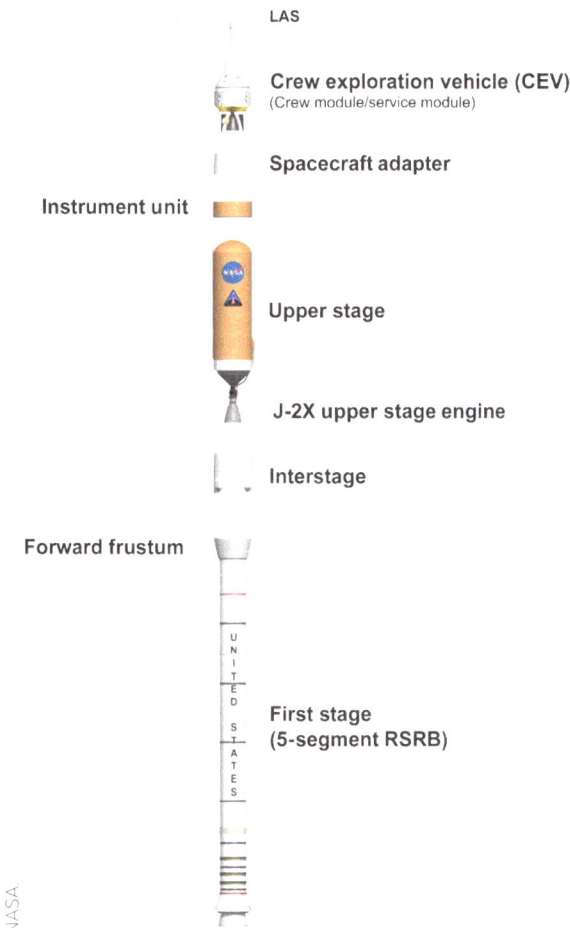

LAS

Crew exploration vehicle (CEV)
(Crew module/service module)

Spacecraft adapter

Instrument unit

Upper stage

J-2X upper stage engine

Interstage

Forward frustum

**First stage
(5-segment RSRB)**

NASA

The PDA process begins with identifying high-risk failure scenarios. This initial step is typically a collaborative effort involving multiple teams from within the Ares I Crew Safety and Reliability (CSR) division. These teams include those responsible for PRA, FMEA, system safety, system analysis and risk assessments (SARAs), and the Ares integrated abort analysis, who work together to discuss and prioritize the assessment cases. Disciplinary experts are also consulted to provide insights into the design properties, potential causes, and effects of the failures:

- Feasibility study: For each selected failure scenario, PDA analysts conduct a feasibility study to evaluate the available data, analysis approach, resources, and scheduling. Once a PDA case is deemed feasible, it is formally formulated with a clear problem statement, data sources, analysis approach, target customers, and a proposed analysis schedule. This formulation ensures that all necessary aspects of the analysis are well-defined and documented.

- Model development and data collection: A PDA model is then developed to describe the physics and system response to the considered failure scenario, including a failure criterion. Input data for the model are obtained from the element office or disciplinary branches. Further consultations are conducted to determine key model parameters and associated uncertainties. The PDA model makes use of Monte Carlo simulations to compute the failure probability. This involves generating random parameter values to evaluate the likelihood of failure.

- Sensitivity analysis: In addition to computing the failure probability, sensitivity analyses are conducted to assess how systematic changes in chosen sensitivity parameters affect the failure probability. This step helps identify which parameters have the most significant impact on the overall risk and informs design adjustments to mitigate these risks effectively.

- Documentation and reporting: Upon completion of the analysis, a detailed PDA report is written to document the study effort and results. These reports are then shared with the CSR teams, Ares I project members, and element offices for inclusion in their respective analyses. For the Ares I LOM and abort-effectiveness assessments, the results are used to update the probabilities of failure scenarios and their consequences and severity.

8.2.1.1.
Example of PDA

Some key components of the Ares I are the Orion capsule, the service module, and the launch abort system. The first stage of the launch vehicle is a reusable solid rocket booster derived from the Space Shuttle, while a J-2X engine powers the upper stage, fueled by liquid oxygen and liquid hydrogen. During ascent, the first-stage booster propels the vehicle for the first two and a half minutes, reaching speeds of about Mach 5.7. After this first stage separates, the J-2X engine ignites to continue propelling the Ares I. An integrated assessment of the Ares I PDR can identify many critical failure scenarios, which can be grouped for clearer analysis and reporting into what are known as "super cut sets," environmental conditions, and failure categories. The overall assessment aims to find out if the Ares I meet the required reliability standards, with a calculated risk below the specified threshold of 1 in 500, as mandated by the System Requirements Document. To refine these assessments, the SIFA team can also conduct multiple PRAs focusing only on high-risk scenarios such as debris during ascent, bird strikes, interstage damage, liftoff umbilical contact, and upper stage engine failures. This is to better quantify probabilities and minimize potential risks to crew safety and mission success.

8.2.1.2.
Upper Stage Engine Uncontained Failure

One of the top risk drivers for LOM and LOC is the uncontained failure of the upper stage engine during boost. A critical cause of such failure is the penetration of the fuel turbopump (FTP) turbine blade fragments during J-2X engine operations. This failure mode, classified as high criticality in the Ares I FMEA, can lead to significant damage if the high-energy blade fragments impact other engine components or

the upper stage. The function of the FTP turbine is to convert the energy from the gas-generator exhaust gas flow into mechanical energy to drive the FTP pump, which raises propellant pressure to meet engine inlet conditions necessary for combustion. Despite being designed to meet structural safety factors, turbine blades can fail due to unexpected operating environments, material defects, or anomalies in the casting process. While the turbine housing is designed to withstand operating conditions, it is not specifically designed to contain blade fragments.

A PDA model for the J-2X FTP turbine blade penetration can use simple equations to see if blade pieces can go through the turbine housing. Various assumptions can be made at this point if the blade fragments or pieces travel straight at the housing wall without considering things like how they move, change shape, or break apart. The model is expected to check if the material of the housing can absorb enough energy to stop the blade pieces based on how fast they are moving. The analysis needs to focus on understanding how the design parameters and uncertainties influence the probability of penetration of the turbine housing by blade fragments. The steps to be taken include:

- Define the problem and objective:
  - Problem: Assess the probability of penetration of the turbine housing by blade fragments during J-2X engine operations.
  - Objective: Quantify the risk of uncontained failure of the upper stage engine and evaluate the impact on the overall system reliability.
- Identify key parameters and variables:
  - Design parameters:
    - Turbine blade material properties (e.g., tensile strength, fracture toughness),

turbine housing material properties (e.g., yield strength, ductility), blade geometry, housing geometry.

- Operational parameters: Blade rotational speed, gas-generator exhaust gas flow conditions.

- Uncertainties: Material defects, operating environment variations, anomalies in the casting process.

- Develop a simplified analytical model:

  - Assumptions: Blade fragments travel straight at the housing wall without deformation or breaking apart.

  - Equations: Use energy absorption and impact mechanics equations to determine if the turbine housing can absorb the kinetic energy of the blade fragments, for example, $E_{kinetic} = \frac{1}{2}mv^2$, where m is the mass of the blade fragment and v is the velocity of the fragment.

- Perform a sensitivity analysis:

  - Determine the most influential parameters affecting the penetration probability by varying one parameter at a time and observing the changes in the output.

- Conduct Monte Carlo simulations:

  - Generate random samples for each uncertain parameter based on their probability distributions (e.g., normal distribution for material properties, uniform distribution for operating conditions). For each set of random samples, calculate the probability of penetration using the analytical model. Run many simulations (e.g., 10,000 iterations) to obtain a distribution of penetration probabilities.

- Analyze the result:

  - Probability distribution: Analyze the output distribution to understand the likelihood of turbine housing penetration.

  - Critical scenarios: Identify scenarios with the highest risk of penetration and their corresponding parameter values.

  - Impact assessment: Evaluate the potential impact on the upper stage structure and nearby engine components.

  For example, suppose the Monte Carlo simulation results show that the probability of penetration is 0.02 (2%) under nominal conditions but increases to 0.1 (10%) under worst-case scenarios. This indicates a significant risk under certain conditions, necessitating design improvements or additional protective measures.

- Develop mitigation strategies:

  - Enhance the turbine housing material or thickness to better absorb the impact energy.

  - Operational adjustments: Modify operational parameters to reduce the likelihood of high-energy blade fragment formation.

  - Inspection and maintenance: Implement more stringent inspection and maintenance protocols to detect and mitigate potential material defects or anomalies.

- Validate the model experimental testing:

  - Conduct physical tests to validate the analytical model and simulation results.

  - Comparison with historical data: Compare the model predictions with historical failure data to ensure consistency and reliability.

The model can assess the probability of blade penetration, showing that the turbine housing might be penetrated in specific situations. Such an approach can suggest if a direct impact on the upper stage structure is improbable or not, or if there are any concerns regarding engine components near the FTP. These results can be used to revise the probabilities of failure scenarios and evaluate how they could affect the structural strength of the upper stage.

To illustrate, consider a simplified model where the blade fragment mass is 0.1 kg and its velocity is 300 m/s. The kinetic energy is: $E_{kinetic} = \frac{1}{2} \times 0.1 \times 300^2 = 4500$ J. Assuming the turbine housing material can absorb up to 5000 J of impact energy without penetration, the initial analysis suggests that penetration is unlikely. However, considering uncertainties and variations, Monte Carlo simulations might reveal a higher probability of penetration under certain conditions. This can be demonstrated by using Monte Carlo simulations to generate random samples for the key parameters (i.e., blade mass, velocity, and the turbine housing's absorption capacity in Figure 8.2a) based on predefined mean values (0.1 kg) and standard deviations (0.02 kg). We can use this information to calculate the kinetic energy of the blade fragments. By comparing this energy, we compute the probability of penetration as the proportion of simulations where penetration occurs, incorporating the uncertainty in the material's randomly generated energy absorption capacity. Each simulation either results in penetration (1) or no penetration (0), and the mean of these binary outcomes provides a straightforward measure of the overall likelihood of penetration across the range of possible conditions accounted for by the simulations; for example, more than 10,000 simulations

resulted in the probability of penetration of 37.81%.

We can also carry out a sensitivity analysis by calculating correlation coefficients to identify which parameters most influence the penetration probability. This helps in understanding the impact of uncertainties on the system's reliability and safety. Figure 8.2b shows the scatter plots to visually analyze the sensitivity of the penetration outcome to the variations in the three key parameters: mass, velocity, and absorption capacity:

- Penetration (1): If the outcome of a simulation is 1, it means penetration occurred. This implies that the kinetic energy of the blade fragment exceeded the absorption capacity of the turbine housing. A penetration outcome indicates a failure scenario where the blade fragment could potentially cause damage to other engine components or the upper stage.

- No penetration (0): If the outcome of a simulation is 0, it means no penetration occurred. This implies that the kinetic energy of the blade fragment was within the absorption capacity of the turbine housing, preventing the fragment from penetrating through the housing. A no penetration outcome indicates a safe scenario where the turbine housing successfully contains the blade fragments, avoiding further damage.

If the correlation coefficient between the blade fragment's velocity and the penetration outcome is high, it indicates that controlling the fragment's velocity is critical in reducing penetration risk. Such a visual inspection can provide additional insights into which parameters are most influential in determining the penetration probability and mitigate the risk of turbine blade penetration effectively.

Figure 8.2 (a) Histograms of the uncertain parameters, (b) scatter plots for sensitivity analysis: penetration
(1 = yes, 0 = no).

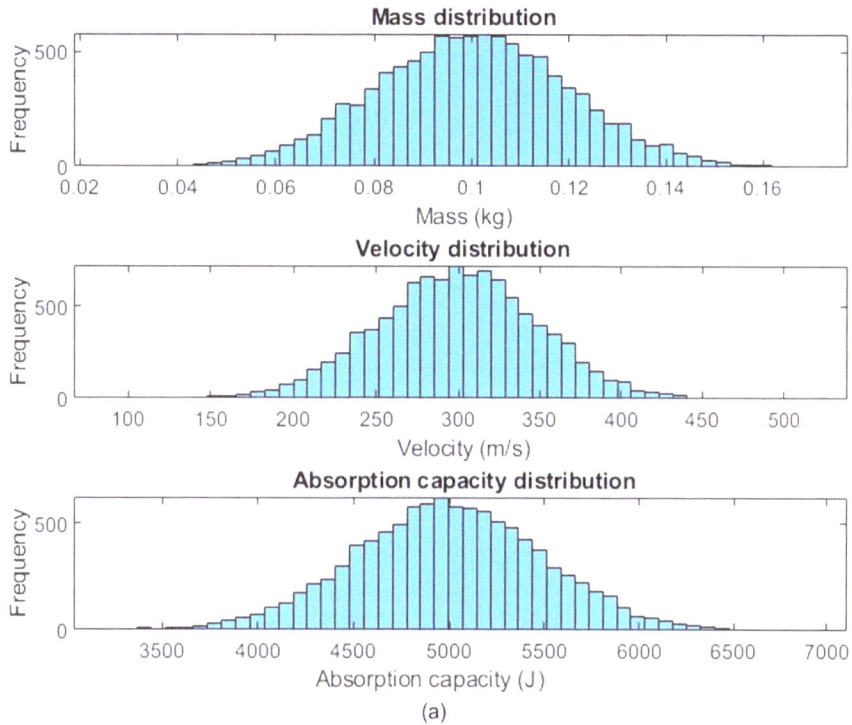

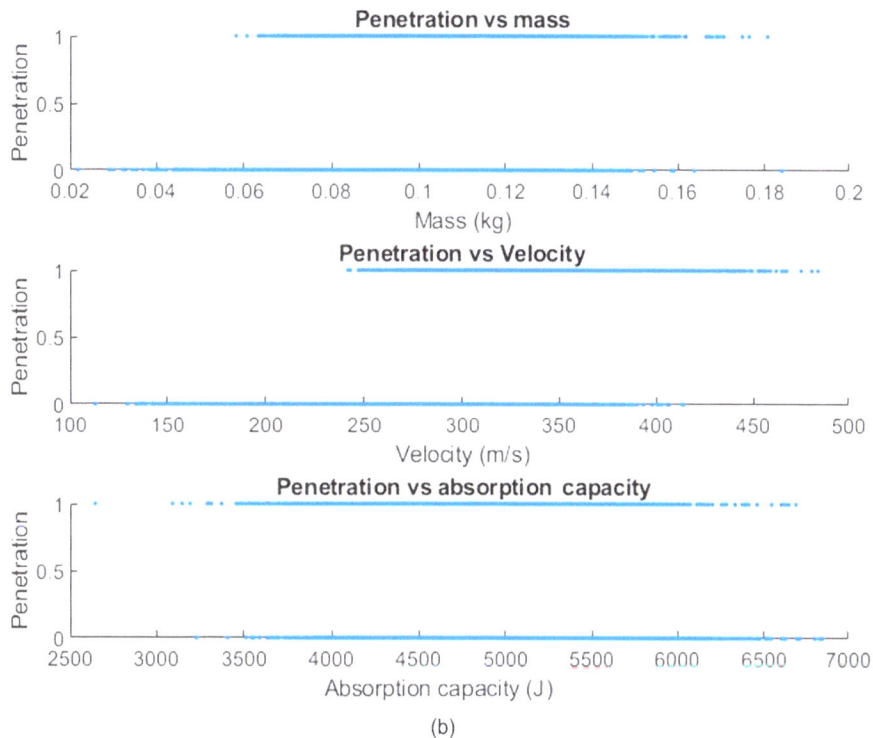

(a)

(b)

8.3.
PRA

PRA is a systematic and comprehensive method used to identify and quantify risks in human spaceflight [8.1]. This method is essential for understanding the likelihood of different failure modes and their potential impact on the overall mission. PRA is widely used because of its ability to provide a quantitative basis for decision-making under uncertainty by evaluating various risk factors, including HW failures, medical causes, environmental conditions, SW reliability, and crew performance. Programs like the ISS

integrate PRA from design through operations to support decision-making. PRA originated in the early 1960s to evaluate safety in designing and operating Intercontinental Ballistic Missiles and later found value in other industries. NASA's initial attempt at reliability analysis during the Apollo program highlighted the high-risk nature of space missions. Over time, PRA became integral to nuclear safety following the Three Mile Island accident, leading to widespread adoption in evaluating and improving plant designs. Table 8.1 summarizes some innovations and benefits of using PRA across various space programs.

Table 8.1 Key risk management and benefits across major space programs.

Space program	Scope	Benefits	Source
Space Shuttle program	• Identified top risk drivers such as TPS damage, leading to on-orbit inspections and repair capabilities. • Risk reduction from 1 in 10 to 1 in 90 by program's end.	• Comprehensive risk identification: enabled NASA to pinpoint critical risk areas like TPS damage. • Cost-effective design changes: early identification of risks allowed for less costly adjustments. • Mission risk reduction: continuous updates to PRA improved mission safety and reduced risk over time.	Pate-Cornell and Dillon [8.7]
ISS	• Assessed risks related to HW, medical emergencies, and micrometeoroid and orbital debris impact. • Supported decision-making for emergency crew return.	• Comprehensive scope: included HW, medical, and environmental risks. • Emergency preparedness: evaluated space vehicle risks for emergency crew return options. • Improved safety protocols: continuous PRA updates enhanced safety measures and preparedness for critical failures.	Duncan [8.8], Heard and Vitali [8.9]
Commercial Crew Program (CCP)	• Ensured LOC and LOM requirements are met for safe crew transportation. • Evaluated risks from launch through docking and return.	• Verification of requirements: PRAs ensured that spacecraft met stringent safety standards. • Comparative studies: provided detailed risk evaluations for different mission phases, helping to choose safer design options. • Enhanced crew safety: regular PRA updates contributed to improved safety protocols for crewed missions.	Boyer et al. [8.10]
Space Launch Systems (SLS)	• Predicted system risk probabilities and diagnosed key risk causes. • Reduced uncertainty in risk models.	• Risk analysis framework: developed a robust framework for analyzing risks using Bayesian networks and fuzzy methods. • Key risk identification: diagnosed critical risk causes and determined risk conduction paths. • Sensitivity analysis: performed detailed sensitivity analyses to address uncertainties and improve reliability.	Zhang et al. [8.11]

© SAE International

The steps to carry out a PRA are as follows:

1. System definition: The first step involves defining the system under consideration, identifying all components, subsystems, and their interactions within the spacecraft. Engineers frequently create detailed diagrams and models to accurately represent the system architecture. For example, the Space Shuttle system included the orbiter, solid rocket boosters, external tank, and various subsystems such as propulsion, communication, and TPSs. Detailed models are created to represent these components and their interactions.

The Space Shuttle program benefited greatly from this step as it provided a comprehensive understanding of the complex interactions within the shuttle system, leading to improved safety protocols.

2. Hazard identification: Pinpoint all potential hazards that could lead to system failures. Techniques commonly used to explore potential failure modes include FMEA and Hazard and Operability (HAZOP) study. For example, potential hazards identified included failures in the TPS, propulsion failures, structural failures, and communication system breakdowns.

The ISS program benefited significantly from this step by identifying various potential hazards, including micrometeoroid and orbital debris impact, enhancing the station's resilience.

In failure mode analysis, we examine specific failure modes for each identified hazard to determine how each component can fail, the causes of these failures, and the resulting consequences. This step often requires extensive historical data and expert judgment.

Then, experts use tools such as event tree analysis (ETA) and FTA. ETA starts with an initiating event and explores possible outcomes based on system responses, while FTA begins with a top-level system failure and works backward to identify root causes. For example, the TPS's failure modes included damage to the heat-resistant tiles and foam shedding from the external tank. The causes and potential consequences of these failures were analyzed using tools like FTA.

The Space Shuttle program extensively used FTA and ETA to understand and mitigate critical failure modes, notably improving the reliability of the TPS.

3. Probability quantification: Once we understand the failure modes and their interactions, the next step is to quantify the probabilities of these events. This involves using statistical data, reliability models, and expert elicitation to assign probabilities to different failure scenarios. For example, probabilities are assigned to various failure scenarios using historical data from previous flights, reliability models, and expert elicitation. For instance, researchers may use data from prior missions to quantify the probability of foam shedding.

The SLS benefited from advanced probabilistic models to quantify the likelihood of critical failures, aiding in the development of robust safety measures.

4. Consequence analysis: Evaluate the impact of failure modes on the overall mission, considering both immediate effects and long-term consequences, including safety, mission success, and financial implications. This step often involves simulations and scenario analysis. For example, the evaluation takes into account the impact of these failures on the mission, including potential damage to the orbiter during reentry, which could cause mission failure or LOC.

The CCP benefited significantly from consequence analysis, ensuring the safety and success of crewed missions by evaluating the impact of potential failures comprehensively.

5. Risk profile integration: The final step integrates the probabilities and consequences of all identified failure modes to provide an overall risk profile for the system. This involves calculating metrics, such as the expected number of failures, system reliability, and risk to mission success. For example, by integrating the results, the team created an overall risk profile, which helped prioritize risks and guide mitigation strategies.

The ISS program used integrated risk profiles to prioritize risks and enhance the station's safety and operational efficiency over its extended mission duration.

8.3.1.
Example of PRA

8.3.1.1.
Liftoff Recontact

The liftoff recontact failure bin was identified as a high LOM risk driver due to conservative estimates based on historical launch data and pre-PDR design limitations. The Ares I liftoff clearance requirement mandates sufficient clearance between the launch vehicle and the launch facility. This scenario spans multiple projects, including the Ares vehicle, Orion crew vehicle, launch tower, mission systems, and natural environments.

To address the risk of recontact during liftoff, a PDA can be conducted to evaluate the potential for liftoff recontact under various conditions

such as wind gusts, thrust vector control (TVC) failures, and nominal ground-support system release. The specific example presented here will focus on the vehicle damping system designed to reduce lateral motion caused by wind-induced oscillations. The following are the steps to be taken:

- Define the system and identify failure modes: System components: Ares I launch vehicle, Orion crew vehicle, launch tower, ground-support systems, environmental factors (wind, weather conditions). Failure modes: Wind gusts causing lateral oscillations, TVC failures, ground-support system release anomalies, and damping system failures.

- Develop an FTA: Create a fault tree that visually represents the various failure modes leading to liftoff recontact. Identify the top event (recontact during liftoff) and break it down into contributing events (e.g., wind gust-induced oscillations, TVC failure, damping system failure).

- Quantify probabilities of basic events: Use historical data, expert judgment, and simulation results to estimate the probabilities of basic events such as: probability of significant wind gusts during liftoff, probability of TVC failure during liftoff, probability of damping system malfunction. Perform Monte Carlo simulations or other statistical methods to generate probability distributions for these events.

- Analyze the impact of wind-induced oscillations: Model the vehicle damping system's performance under varying wind conditions. Evaluate how effective the damping system is in reducing lateral motion caused by wind gusts. Assess the residual risk of recontact after considering the damping system's mitigation effect.

- Evaluate system interdependencies: Analyze the interdependencies between different systems (e.g., how TVC failure might affect the damping system's performance). Consider how environmental factors interact with mechanical and control systems during liftoff.

- Calculate the overall risk of recontact: Combine the probabilities of individual failure events using the FTA. Calculate the overall probability of recontact during liftoff. Use risk metrics such as the expected value of losses, the probability of LOM, and confidence intervals to quantify the risk.

- Risk mitigation and decision-making: Identify and prioritize risk mitigation measures based on the PRA results. Consider design improvements, additional testing, or operational changes to reduce the risk of recontact. Make informed decisions regarding the implementation of these measures to enhance the safety and reliability of the launch.

We can perform FTA to identify the possible failure modes leading to liftoff recontact and then use ETA to evaluate the sequences of events following an initial failure. A simple fault tree can be defined where the top event is liftoff recontact:

- Wind gusts causing significant lateral displacement.
- TVC failure.
- Damping system failure.

Using Monte Carlo simulations, we can generate random wind speeds and calculate the resulting lateral displacement, considering possible TVC and damping system failures. We can then check if the displacement exceeds the recontact threshold to determine recontact occurrences to calculate the overall probability of recontact.

We can also construct an event tree to identify possible sequences of events following significant wind gusts and their associated probabilities. This comprehensive approach provides a detailed and accurate assessment of the risk of liftoff recontact.

Similar to the previous example, we can run Monte Carlo simulations to model the effects of uncertainties in wind gust speed and damper release time. For each simulation, random values are generated for wind gust and damper release time based on their respective mean values (10 m/s) and standard deviations (3 m/s). This uncertainty distribution is shown in Figure 8.3a. The lateral displacement caused by wind can then be calculated using a simplified linear model to determine if recontact occurs. The overall probability of recontact is computed as the mean of the outcomes across all simulations; for example, for more than 10,000 simulations, we computed the:

- Probability of recontact: 2.68%.
- Mean lateral displacement with damping: 0.2245.
- Standard deviation of lateral displacement with damping: 0.1783.
- Correlation between wind speed and lateral displacement: 0.3804.

This information will be useful during mitigation efforts and help prioritize resources. In addition, a sensitivity analysis can be performed to understand the correlation between the parameters (wind gust and damper release time) and the recontact outcome, identifying the most influential factors. These relationships of the parameters with recontact outcomes are shown in Figure 8.3b, which provides some insights into how variations in these parameters affect the probability of recontact:

Figure 8.3 (a) Distribution of wind speeds and lateral displacement, (b) recontact occurrences.

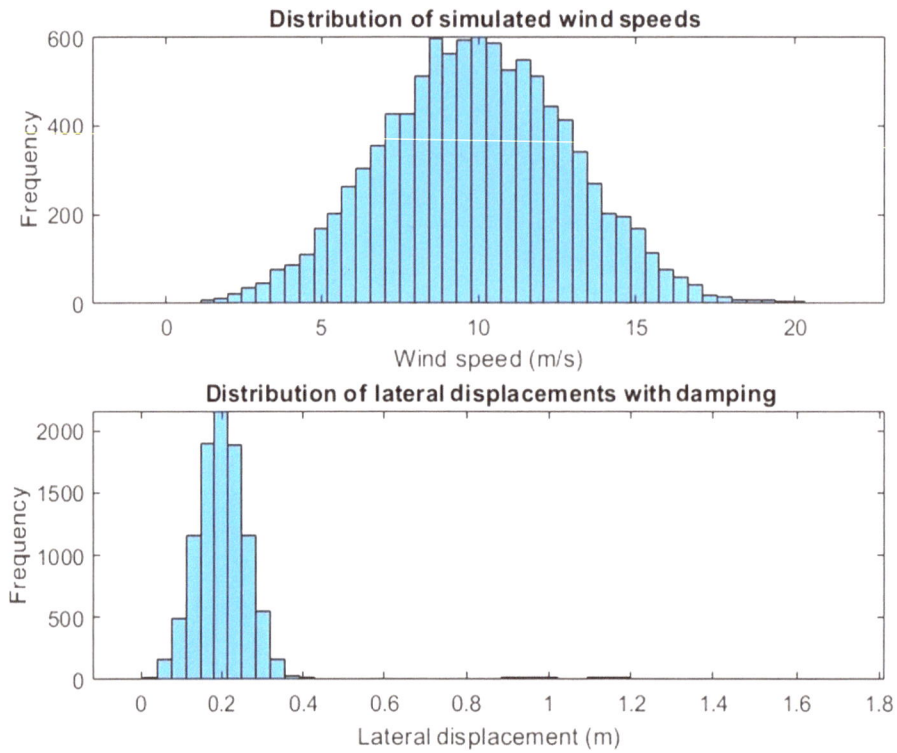

(a)

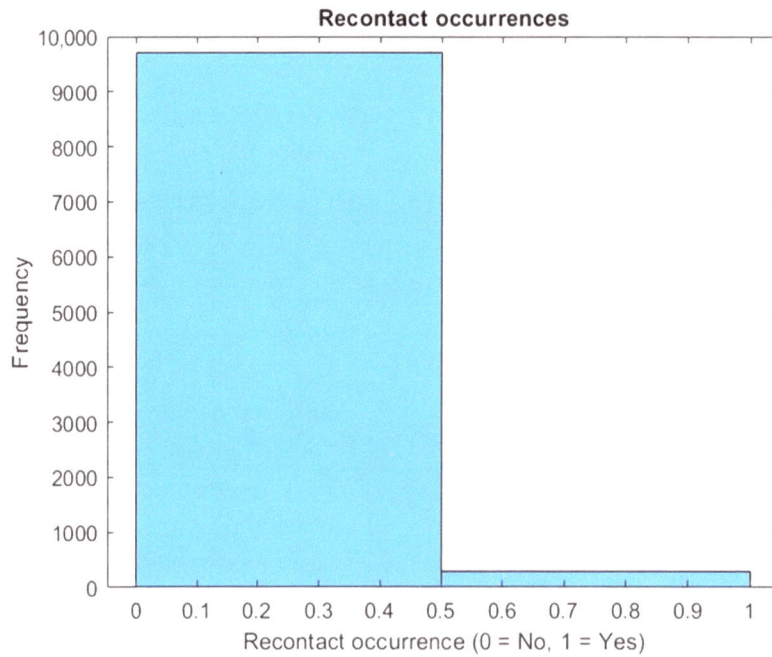

(b)

- Recontact (1): If the outcome of a simulation is 1, it means recontact occurred. This implies that the lateral displacement of the launch vehicle, caused by wind gusts and the timing of damper release, exceeded the clearance threshold (0.5 m in this example). A recontact outcome indicates the failure scenario where the vehicle may collide with the launch tower or other structures during liftoff, posing a high risk to the mission.

- No recontact (0): If the outcome of a simulation is 0, it means no recontact occurred. This implies that the lateral displacement was within the clearance threshold, and the launch vehicle did not collide with any structures during liftoff. A no recontact outcome indicates a safe scenario where the vehicle maintains sufficient clearance and successfully avoids any collision during liftoff.

In this PRA example, FTA and ETA play an important role in systematically evaluating the risks associated with liftoff recontact during a rocket launch. FTA helps in identifying and visualizing the various failure paths that could lead to a liftoff recontact by breaking down the top event into intermediate and basic events (Figure 8.4a). This breakdown allows us to quantify the probabilities of individual failures such as TVC failure, damping system failure, and significant wind gusts. By understanding these probabilities, we can determine the likelihood of the overall risk event. On the other hand, the ETA complements FTA by evaluating the sequences of events following an initial failure, such as significant wind gusts. It maps out different scenarios and their possible outcomes (Figure 8.4b), helping us understand how initial failures propagate through the system; for example, the ETA shows the paths where TVC or damping system failures could lead to recontact, considering different combinations of these failures.

Figure 8.4 (a) FTA identifying events that could lead to failure, (b) ETA showing different scenarios and their possible outcomes.

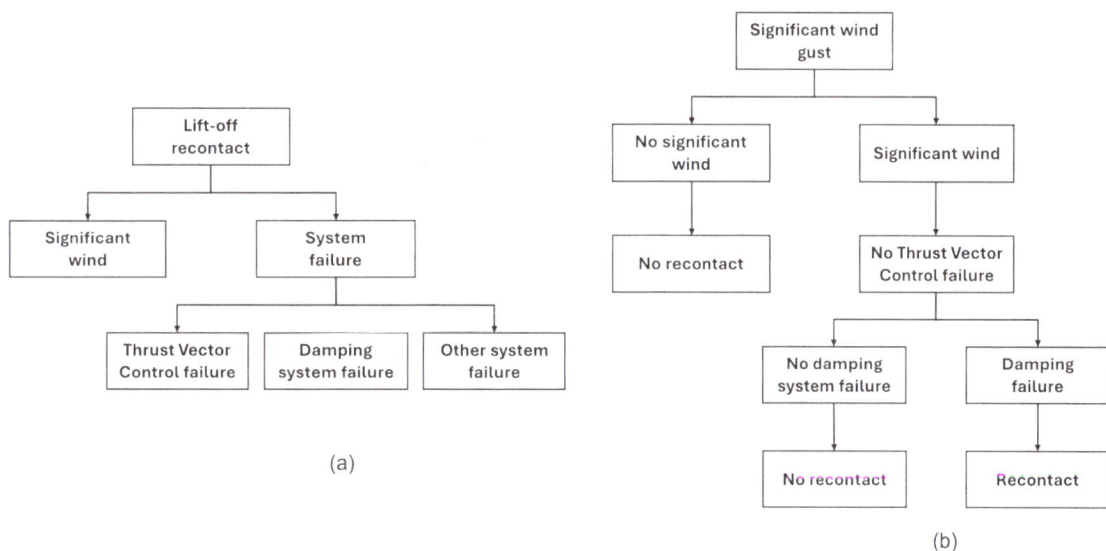

© SAE International

(a)

(b)

Such a structured approach can guide decision-making and mitigation efforts, ensuring that we focus on the most significant risks and their pathways. Integrating FTA and ETA into Monte Carlo simulations also helps the PRA provide a robust and realistic assessment of the likelihood and consequences of liftoff recontact.

8.4.
Other Approaches

8.4.1.
HA

HA is a systematic process used to identify potential hazards and assess their risks in complex systems. This technique is essential in the aerospace industry to ensure the safety and reliability of spacecraft. HA aims to pinpoint possible dangers early in the design phase, allowing engineers to mitigate risks before they manifest during a mission. By identifying and analyzing hazards, HA helps in developing strategies to prevent accidents and failures, enhancing the overall safety of space missions. One notable application of HA in the space industry is its use in NASA's Mars rover missions. The Mars rover missions involve highly complex systems designed to explore the Martian surface, requiring meticulous hazard identification and mitigation to ensure mission success.

8.4.2.
FMEA

FMEA is a structured approach to identifying potential failure modes within a system, understanding their causes and effects, and prioritizing the associated risks. This technique is widely used in the aerospace industry to enhance the reliability and safety of spacecraft by systematically analyzing components and subsystems for potential points of failure. FMEA helps engineers to identify critical areas that require design improvements or additional safeguards, ensuring that missions can proceed with minimal risk of unforeseen failures. The ISS is a prime example of the application of FMEA in the aerospace industry. Given the complexity and critical nature of the ISS, FMEA has been an essential tool for ensuring the station's safety and reliability.

A summary of these techniques, their strengths, and weaknesses are shown in Table 8.2.

Table 8.2 Strengths and weaknesses of common safety and risk assessment techniques.

Technique	Strengths	Weaknesses
HA	Identifies potential hazards early in the design process	Can be subjective and dependent on the experience of the analyst
	Provides a broad overview of safety issues	May not identify all possible hazards—simple and straightforward to perform
	Simple and straightforward to perform	Does not quantify the likelihood or impact of hazards
FMEA	Systematically identifies potential failure modes and their effects	Time-consuming and labor-intensive
	Prioritizes failure modes based on severity, occurrence, and detection	Can be overly detailed, leading to analysis paralysis
	Encourages detailed understanding of system components	May miss complex interactions between failure modes
FTA	Provides a clear visual representation of the logical relationships between failures	Can be complex and difficult to construct for large systems
	Helps identify root causes of failures	Helps identify root causes of failures
	Quantifies the probability of top-level events based on lower-level events	Does not consider the timing or sequence of events
PRA	Quantifies the likelihood and consequences of failure scenarios	Requires extensive data and computational resources
	Provides a comprehensive risk profile	Can be complex and difficult to interpret without specialized knowledge
	Facilitates objective comparison and prioritization of risks	May involve significant uncertainty in probability estimates
PDA	Provides realistic analysis by treating variables as probability distributions	Requires detailed data and computational resources
	Identifies parameter combinations leading to higher failure probabilities	Can be complex and computationally intensive
	Facilitates design improvements by avoiding high-risk parameter combinations	Dependent on accurate modeling of underlying physics
	Enhances existing techniques like PRA with detailed failure probability quantification	May involve significant uncertainty in model assumptions and parameter estimates

8.5. Summary

Probabilistic analysis in reliability engineering remains essential for the development and deployment of spacecraft. They allow engineers to account for uncertainties in material properties, environmental conditions, and operational stresses, leading to more robust and reliable systems. The tools and techniques discussed in this chapter, such as HA, FMEA, and probabilistic assessments, offer comprehensive frameworks for identifying, quantifying, and mitigating risks. These methods enable a deeper understanding of potential failure modes and their impacts, facilitating proactive risk management and design optimization. For instance, the SIFA on the Ares I launch vehicle demonstrates how probabilistic methods can enhance reliability and safety in space missions. The process of developing and implementing PDA models highlights the importance of early risk identification and continuous improvement in design and operations. Moreover, reliability prediction of spacecraft using probabilistic models underscores the necessity of anticipating and managing potential failures over time.

As health management technology advances and missions become more ambitious, the role of probabilistic methods in reliability engineering will only continue to grow.

References

8.1. NASA, "Probabilistic Risk Assessment Procedures Guide for NASA Managers and Practitioners: A Comprehensive Guide on PRA Methodologies," NASA/SP-2011-3421, 2011.

8.2. Birolini, A., *Reliability Engineering: Theory and Practice*, 8th ed. (Berlin: Springer, 2017).

8.3. Modarres, M., *Reliability and Risk Analysis: A Guide for Practitioners* (Boca Raton: CRC Press, 2018).

8.4. Calhoun, J., Lin, L., and Young, G., "Human Reliability Analysis in Spaceflight Applications," in Carayon, P. (ed.), *Handbook of Human Factors and Ergonomics in Health Care and Patient Safety*, 2nd ed. (Boca Raton: CRC Press, 2015).

8.5. Milovanov, V.A., "Reliability Prediction for Spacecraft," AIAA, 2008.

8.6. Yang, H., Zio, E., and Zeng, D., "An Accelerated Simulation Approach for Multistate System Mission Reliability and Success Probability under Complex Mission," *Reliability Engineering & System Safety* 155 (2016): 114-124, doi:https://doi.org/10.1016/j.ress.2016.06.005.

8.7. Pate-Cornell, E. and Dillon, R., "Probabilistic Risk Analysis for the NASA Space Shuttle: A Brief History and Current Work," *Reliability Engineering and System Safety* 74, no. 3 (2001): 345-352, doi:https://doi:10.1016/S0951-8320(01)00081-3.

8.8. Duncan, G., "International Space Station End-of-Life Probabilistic Risk Assessment," NASA Technical Reports Server, 2014.

8.9. Heard, A. and Vitali, R., "The International Space Station Probabilistic Risk Assessment Fire Analysis, Sensitivity Studies for Critical Variables, and Necessary Areas of Additional Development," in Spitzer, C., Schmocker, U., and Dang, V.N. (eds), *Probabilistic Safety Assessment and Management* (London: Springer, 2004), doi:https://doi.org/10.1007/978-0-85729-410-4_192.

8.10. Boyer, R.L., Hamlin, T.L., Grant, W.C., Stewart, M.A. et al., "'Making Safety Happen' through Probabilistic Risk Assessment at NASA," 2019, Retrieved from NASA Technical Reports Server.

8.11. Zhang, H., Wang, Y., and Li, X., "Probabilistic Risk Assessment in Space Launches Using Bayesian Network with Fuzzy Method," *Aerospace* 9, no. 6 (2022): 311, doi:https://doi.org/10.3390/aerospace9060311.

The discussions presented in the previous chapters show the overwhelming interest in achieving autonomous capability; it is likely to influence the adoption of onboard processing during missions. As data-driven solutions become more prevalent, this demand will probably continue to increase. Using onboard health monitoring algorithms on spacecraft can help maintain reliability by managing critical functions and redundancy configurations. However, this is not without its challenges. The space environment is unique and natural phenomena may have significant implications because of radiation exposure and temperature fluctuations. Space-qualified devices face unique challenges and require specialized design considerations to ensure functionality and reliability in harsh space environments. Limited validation and verification standards present a significant gap [9.1]. The design and capabilities of installed onboard processing systems may limit power and memory optimization needs.

This chapter examines the evolution of onboard deployment and the challenges it presents. It provides an in-depth understanding of onboard architectures, starting with an examination of avionics options for health management. It covers their historical development and evaluates their strengths and weaknesses. Next, we examine integrating functions into spacecraft SW. Finally, the chapter discusses various architectural approaches and outlines the primary verification techniques for onboard processing health management solutions.

9.1.
Onboard Decision-Making

Commercialization in the space industry has resulted in a rise in satellite launches and a shift toward fleets of smaller satellites. This has led to a higher acceptance of risk-taking and a willingness to replace faulty ones. Cheaper options for development and manufacture, such as the use of COTS components, have also become more prevalent, resulting in shorter development times and greater modularity. While promising, these changes have also brought new challenges, including a greater need for skilled ground operators. One solution to address their issue is by making real-time decisions on board the spacecraft.

Not only would this make the assets more reliable, efficient, and safer, but it could also help us respond to unforeseen situations without waiting for ground intervention. This would also cut down on mission costs by reducing our dependence on costly ground operators. If we can preprocess much of the telemetry data onboard, we can prioritize relevant information before taking any downlink actions. This means less strain on limited bandwidth and ground operator work and additional time for us to focus on analyzing the most important data. The potential to reconfigure spacecraft parameters in real-time for missions and run algorithms across multiple satellites is incredibly exciting. It is a whole new paradigm that could solve the current many challenges and open various possibilities for deep space exploration.

Two aspects need to be discussed. The OBCs are critical for spacecraft function control. Depending on the initial design, these will require redundancy, electromagnetic compatibility, and radiation robustness. The other aspect is the onboard SW (OBSW) that will manage the spacecraft's functions. Depending on the service architecture, the SW will ensure real-time control, handle data input/output (I/O), and the management of Attitude and Orbit Control System (AOCS), thermal, power, and health [9.2].

Modern spacecraft are composed of various subsystems, including thermal control electronics, electrical power systems, mechanisms, telecommunications, and central processing. Onboard buses and networks interconnect all these elements. Within this intricate onboard system architecture, a subsystem dedicated to "high-performance computing" is essential for future space missions involving tasks such as active debris removal, precision landing, or rendezvous operations. This subsystem handles the real-time processing of inputs from spacecraft sensors, using specialized health management SW or HW units.

The future of OBCs is increasingly leaning toward supporting sophisticated and healthy functions for spacecraft. These involve continuous monitoring and analysis of various subsystems to ensure optimal performance and identify potential failures. This task is demanding in computational resources, especially when it includes complex sensor data analysis and predictive maintenance algorithms.

To meet these demands, one must handpick the HW for processors or coprocessors. We can categorize the processing devices into several types: general-purpose processors/microcontrollers, digital signal processors (DSPs), graphical processing units (GPUs), and FPGAs [9.3]. In addition to these general categories, manufacturers offer specialized HW options such as image signal processors, AI engines, and neural processing units. These specialized options are significant for specific SHM tasks, offering enhanced capabilities for real-time health monitoring and predictive maintenance (Figure 9.1 and Table 9.1).

Figure 9.1 Onboard processing architectures.

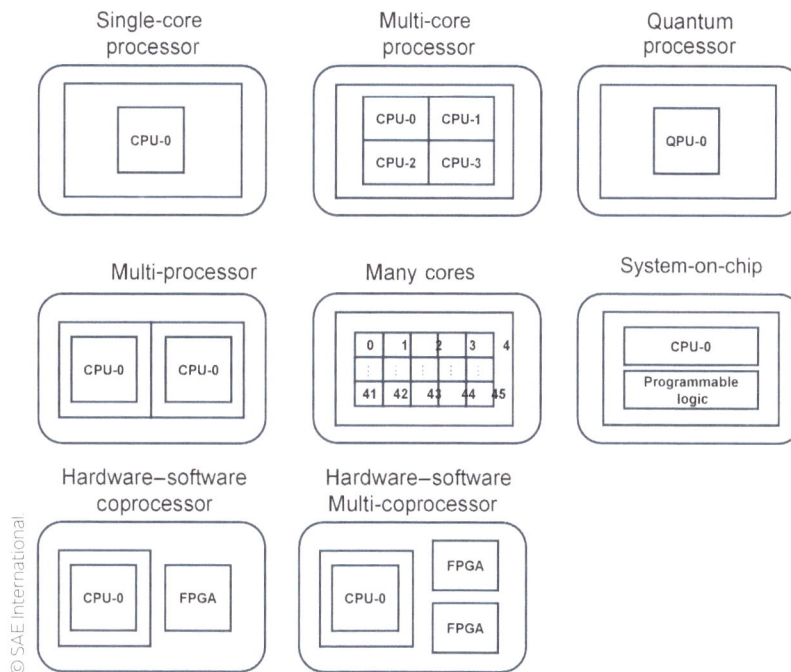

© SAE International

Table 9.1 Various processor types in the market.

Component type	Description
General-purpose processor/ microcontroller	• Common processor with predefined chip architecture, executing sequential programs. Includes arithmetic units, internal registers, buses, and memory access. Used in computers and mobile phones. Operates with or without an operating system (OS). Supports multiprocessor and multicore configurations. • It acts as the CPU in most spacecraft and satellite systems, handling a range of onboard system tasks, including command and control, data handling (DH), and running GNC applications.
DSP	• Microprocessor designed for processing digital signals in real time. Ideal for signals like video, audio, and images. Capable of complex mathematical operations in one clock cycle. Often used as a coprocessor for specific functions like convolutions and filters. • It is used in signal processing tasks that require real-time data processing, such as handling telemetry data, sensor signal processing, and communications.
GPU	• Specialized microprocessor for image and video processing. Contains multiple parallel processing units. Known for high power consumption but efficient for handling large data sets in image processing. • Advanced imaging tasks, including EO and deep space imaging, employ it. Users can employ GPUs for high-data-volume processing tasks, such as image analysis and AI applications.
FPGA	• Programmable and reconfigurable digital processing component. Comprises a matrix of logic gates for implementing various functions. This component allows for the programming of the HW itself, unlike processors that execute SW on predefined HW. Used for creating prototypes and implementing fault-tolerant processors. • People use FPGAs for custom digital circuit development, allowing them to create flexible and reconfigurable system designs. FPGAs are a popular choice for prototyping, implementing specific control logic, and fault detection.
Application-specific integrated circuit (ASIC)	• Dedicated HW design for a specific function, more specific than FPGA. ASICs are not programmable or reconfigurable after manufacturing. Commonly used in processor chip creation. • Engineers use ASICs when they need a fixed, optimized, and efficient HW solution for specific tasks in spacecraft systems, such as dedicated sensor processing or specific control algorithms.
System-on-chip (SoC)	• Integrates multiple HW components (processor, FPGA, DSP, etc.) into a single device. Reduces system power and offers fast communication between components. Supports a variety of accelerators and enhances flexibility during implementation. • Challenging to produce in space format. • SoCs provide integrated solutions combining processing power and specialized functionality, leading to more efficient and compact systems suitable for space applications. They offer a balance of power, flexibility, and size, necessary for satellite technology.

9.2.
Evolution of Onboard Processing

In the early days of spacecraft design, such as the NASA Mercury program, the focus was on assessing whether humans could survive in space under zero gravity conditions. The programs used chimpanzees as "crew" for initial flights, with spacecraft control performed from the ground. The later missions allowed manual control by astronauts, marking a transition toward more advanced spacecraft control techniques. In subsequent programs, like Gemini, engineers developed and qualified essential technologies, including OBCs, to enable more complex missions. These spacecraft showcased capabilities such as environmental control, life support systems, rendezvous maneuvers, and docking maneuvers, thanks to the engineers equipping them with digital computers. Advanced onboard computing capabilities based on ICs facilitated GNC, representing further breakthroughs made by researchers. The SW development team used assembly language to write the SW, incorporating features such as real-time multitasking functions, redundancy, and advanced voting systems.

Several advancements in onboard computing capability were also made. The introduction of transistor-based digital sequencers and improved core memories marked the beginning of an era where programmable commands were prioritized, targeting mission-specific objectives and adapting to varying computing architectures. The progress in exploring multiple planets highlights developments in ICs, memory technologies, and digital data transmission. Many of these spacecraft used microprocessors and reduced instruction set computer (RISC) architectures, representing a shift toward more powerful and radiation-hardened processors. For, the RAD750, based on IBM PowerPC 750, became a standard for various missions, focusing on reliability and performance. These processors powered various missions, including Mars rovers (Spirit and Opportunity), Mars Reconnaissance Orbiter, and the Spitzer Space Telescope.

The recent transition to SoC technology highlights the integration of CPU and interface controllers into a single chip. For example, today's technology often uses the LEON3FT architecture as it is a prominent example of contemporary onboard computing, emphasizing SoC design. Table 9.2 shows a concise overview of each program, highlighting their unique characteristics and contributions to onboard computing technology. Integrating critical components, such as CPU, interface controllers,

and real-time operating systems (RTOSs), also showcases advanced capabilities in other areas:

- Autonomous control: Manned space missions require spacecraft to operate autonomously in various phases, including launch, orbit, rendezvous, docking, and reentry. OBCs provided the capability for autonomous control, allowing the spacecraft to perform complex maneuvers without constant intervention from ground control.

- Navigation and guidance: OBCs were essential for navigation, guidance, and trajectory calculations. They facilitated precise control of the spacecraft's orientation, velocity, and position, ensuring accurate execution of mission objectives, especially during critical maneuvers such as orbital rendezvous and lunar landings.

- Real-time decision-making: OBCs enable real-time decision-making capabilities. Astronauts could respond to unexpected situations, adjust the mission plan, and take manual control if necessary. This was crucial for ensuring the safety and success of the mission, especially when faced with unforeseen challenges.

- Reducing communication delays: The vast distances between Earth and spacecraft introduced communication delays. OBCs allowed for immediate response to changing conditions without waiting for instructions from ground control. This autonomy was important during critical phases where split-second decisions were required.

- Data processing and analysis: OBCs processed data from various sensors, instruments, and spacecraft systems. They analyzed these data to monitor the spacecraft's health, identify potential issues, and start corrective actions.

- System redundancy and reliability: Manned missions incorporated redundancy in OBC systems to enhance reliability. Engineers incorporated redundant computers and voting systems to enhance reliability, allowing critical functions to continue even in the presence of failures.

- Mission flexibility: OBCs provide the flexibility to adapt to changing mission requirements and conditions. Making real-time adjustments during lunar descent was vital for astronaut safety in missions like Apollo.

- Automation of routine tasks: OBCs automated routine tasks, reducing the workload on astronauts, and allowing them to focus on critical mission activities. This automation enhanced the efficiency of spacecraft operations.

TABLE 9.2 Contributions to onboard processing.

Space program	Importance of onboard computing
Mercury program	• The initial focus was on testing whether humans could survive in space. • Manual control from the ground due to spacecraft limitations. • Later missions allowed manual control by astronauts. • Onboard computing played a minimal role; analog control techniques were used.
Gemini program	• Demonstrated and qualified advanced technologies for lunar missions. • OBCs are essential for tasks like rendezvous, docking, and reentry. • Gemini Digital Computer (GDC) with magnetic core memory. • First use of OBC for nontrivial space maneuvers.
Apollo program	• Onboard computing is crucial for lunar missions (e.g., Apollo 11). • Apollo Guidance Computer (AGC) is used for GNC. • ICs replaced discrete transistors. • Real-time multitasking sublayer and magnetic core memory.
Space Shuttle	• IBM AP-101 computers controlled the Space Shuttle. • Shuttle Data Processing System with redundancy and voting scheme. • Multiple upgrades over the years, from magnetic core to semiconductor memory. • HAL/S high-level language developed to program the AP-101.
Satellite/probes	• Onboard computing is crucial for autonomous control, navigation, and data processing. • Enhanced reliability through redundancy. • Real-time decision-making capability. • Adaptability to changing mission requirements. • Automation of routine tasks to reduce astronaut workload.
Mariner 10 mission	• Simple "Digital Command Sequencers" for onboard control. • Programmable execution of basic commands.
Voyager 1/2 missions	• Redundant OBCs for communication, control, and data processing. • Shift to ICs.
Galileo Jupiter mission	• RCA 1802 microprocessor used for spacecraft control. • Dedicated processors for camera image preprocessing.

9.2.1.
Device Development

The design of space-qualified devices accounts for their ability to operate in extreme conditions like deep vacuum, strong vibrations, and wide temperature ranges. These also need to mitigate electrostatic discharges and rely solely on thermal conduction for heat dissipation. These devices face a significant challenge in handling ionizing radiation in space. It can lead to single-event effects (SEEs) as well as long-term damage like total ionizing dose (TID) and total nonionizing dose. SEEs include temporary disruptions like single-event transients (SETs), SEUs, and single-event functional interrupts (SEFIs), as well as irreversible damage, like single-event latchups (SELs). SETs are logical transients caused by ionizing radiation, while SEUs, having the ability to alter system states, are bitflips in registers and memories due to radiation strikes.

To combat these radiation effects, devices typically employ radiation-hardened by design (RHBD) or radiation-hardened by process (RHBP) techniques. However, these methods are more expensive than standard procedures due to their specialized nature and limited market demand. Additionally, space computing architectures often incorporate HW and information redundancy techniques, such as TMR and error correction codes, to effectively mitigate SEEs. While these techniques enhance reliability, they also increase the complexity and decrease the performance of the designs.

Space processors, compared to their commercial counterparts, are generally slower. This lag is due to the extensive qualification process required for space-grade components and the industry's preference for tried-and-tested technologies. Over time, this has led to a growing performance gap between space and commercial processors. For instance, the radiation-hardened version of the PowerPC 750 showed only a modest improvement over 15 years, compared to a significant advancement in its commercial version.

This means that traditional microprocessors and ASICs are still being used for critical tasks and deep-space missions. These processors are not suitable for processing heavy tasks or running AI algorithms. As a result, there is a need for more powerful processing solutions, particularly for neural networks like CNNs that require dense matrix operations. For example, neural networks such as Mobile-URSONet have demonstrated potential for enhancing onboard spacecraft pose estimation, reducing reliance on ground-based systems [9.4].

Current space-grade data processing systems, such as the GR740, demonstrate significant advancements in space technology. The GR740, highlighted as the ESA Next Generation MicroProcessor, marks a shift from single-core to multicore, high-performance processors in space applications. This development began in 2009, with the prototypes delivered in 2013. The GR740 is an ASIC that features a quad-core LEON4 processor along with various communication interfaces. This includes the IBM PowerPC processors and CAES Gaisler LEON microcontrollers, listed in Table 9.3. The SPARC V8 architecture serves as the foundation for LEON-based processing cores, which have undergone continuous improvements over the past two decades. LEON3 and LEON4, in particular, have introduced fault-tolerant variants. Multicore processors have become increasingly common in recent years. In the short-term, there is a trend of using ARM microprocessors, while in the long-term, the trend is toward processors with RISC-V architecture, such as NOEL-V, which is open-source.

Table 9.3 ASICs found in space applications.

Processor name	Manufacturer	Processor core	Performance (MHz)
RAD750	BAE Systems	PowerPC 750	266
RAD5545	BAE Systems	PowerPC 750	200
LEON3FT	CAES	SPARC V8	80–100
LEON4FT	CAES	SPARC V8	160–200

- BAE Systems RAD750: This is a radiation-hardened version of the PowerPC 750 microprocessor, and it is one of the most widely used processors in space applications. It has a clock speed of up to 200 MHz and can perform up to 266 million instructions per second (MIPS). With an estimated cost of around $200,000 per unit, the RAD750 processor has played a role in various space missions, such as the Mars rovers Curiosity and Perseverance.

- BAE Systems RAD5545: This is a radiation-hardened version of the ARM Cortex-A76 processor, which is commonly used in mobile devices. The RAD5545 has a clock speed of up to 1.5 GHz and can perform up to 4.7 billion instructions per second (BIPS). It is designed to be used in high-performance computing applications, such as image processing and ML. The RAD5545 is expected to cost around $300,000 per unit.

- Despite their limitations, these processors have built-in fault tolerance and enable cheaper manufacture, shorter development times, and potential reuse of multiple applications.

ArianeGroup, PTScientists, and Roscosmos have announced plans to adopt the GR740 for upcoming missions. Meanwhile, many current spacecraft architectures predominantly utilize the LEON3FT processor. The SPARC V8 RISC instruction set architecture (ISA) serves as the foundation for both the LEON3 and LEON4

processors. They selected the SPARC V8 because of its maturity and simplicity, and to avoid the need for cross-compiling, thanks to the availability of SPARC-based workstations. However, the SPARC ISA, despite its initial expectations in the late 1990s, did not gain widespread adoption outside of space applications. This limitation led to the development of products like NOEL-V, which is based on the RISC-V ISA. The RISC-V ISA has sparked considerable interest in various industrial applications due to its open standard and broader applicability. This shift underscores a growing trend in the space industry toward more versatile and widely supported processor architectures.

HW accelerators such as GPUs, FPGAs, and DSPs can allow for implementing complex solutions onboard satellites. These platforms have different architectures and offer an optimized trade-off between model complexity, processing speed, and power consumption. Table 9.4 shows a selection of high-performance AI platforms that can do both floating-point and fixed-point calculations. Vision-based navigation tasks benefit from using GPUs because of their ability for data parallelization, while FPGAs can enhance inference speed. COTS FPGAs are more powerful than space-grade FPGAs but have longer development times. The Nvidia Jetson Xavier NX outperforms other boards in processing power but has higher power consumption.

Table 9.4 Available COTS products.

Name	Manufacturer	Technology	Power consumption
Nvidia Jetson Xavier NX	Nvidia	ARM CPU, Nvidia Volta GPU	15 W
Google Coral Dev Board	Google	ARM CPU, Google Edge TPU	5 W
Intel Movidius Neural Compute Stick 2	Intel	USB stick, Movidius Myriad X VPU	1 W
Raspberry Pi 4 Model B	Raspberry Pi Foundation	ARM CPU, VideoCore VI GPU	3.5 W
OpenVINO Starter Kit	Intel	Intel Core CPU, Intel Arria 10 FPGA	25 W
Coral Dev Board Mini	Google	Edge TPU	5 V DC, 3 A
Zynq UltraScale+ MPSoC	Xilinx	FPGA + ARM	5–20 W
Qualcomm Flight Pro	Qualcomm	GPU + DSP + ARM	5 V DC, 3 A

Space data processing systems often use FPGAs as coprocessors for executing various computationally intensive algorithms. These algorithms include high-speed communication and file delivery protocols, data compression, and signal processing applications.

Unlike COTS products, space-qualified FPGAs use larger technological nodes and specific designs to ensure adequate radiation tolerance. To enhance the radiation resistance of these FPGAs, engineers use different approaches. For instance, Microsemi FPGAs use a flash-based design to increase tolerance to TID and a specific CMOS design for improved SEEs resistance. The latest releases from Microsemi include the PolarFire FPGA and RTG4, which are realized in 28 nm and 65 nm processes, respectively. Another player in this field, NanoXplore, uses a RHBD static random-access memory (SRAM)-based technology process to achieve higher performance. NanoXplore utilizes a 65 nm process for the development of NG-MEDIUM FPGAs and a 28 nm process for the development of NG-LARGE FPGAs. FPGA technology plays a crucial role in space applications, balancing processing power with resistance to harsh space conditions, particularly radiation. Using larger technological nodes and specialized processes in space-qualified FPGAs is a strategic choice to meet these unique requirements.

It is important to note that the performance, power consumption, and cost of these platforms can vary depending on the specific configuration and intended use.

9.2.2.
Examples of ARM SoCs in Space

The integration of COTS ARM SoCs in aerospace applications showcases how consumer technology can be effectively repurposed for space exploration. This includes two notable examples: the Mars Helicopter (Ingenuity) and the Get Away Special Passive Attitude Control Satellite (GASPACS) CubeSat, highlighting the benefits and trends of using ARM SoCs in these missions.

9.2.2.1.
Mars Helicopter (Ingenuity)

The Mars Helicopter, Ingenuity (shown in Figure 9.2), employs a Qualcomm Snapdragon 801 SoC, a processor originally designed for smartphones. This repurposing demonstrates the adaptability of COTS technology for space missions. The Snapdragon 801 provides a balance of high processing power and energy efficiency, essential for Ingenuity's autonomous operations on Mars. Key capabilities include:

- Image processing: Crucial for navigation and capturing high-resolution photos.
- Navigation: Supports Visual-Inertial Odometry (VIO) for precise movement tracking.
- Communication: Facilitates data transmission between the helicopter and the rover.

Figure 9.2 Mars Ingenuity Helicopter.

NASA.

These features enable the helicopter to navigate the Martian terrain autonomously, identifying and reaching distant science targets [9.5]. Additionally, the Snapdragon 801 has been benchmarked for various space applications, including machine vision and hyperspectral compression, proving its versatility and robustness in harsh environments [9.6].

9.2.2.2.
GASPACS CubeSat

The GASPACS CubeSat from Utah State University utilizes a Raspberry Pi, another COTS ARM-based processor (Figure 9.3). CubeSats are increasingly popular due to their cost-effectiveness and flexibility. The Raspberry Pi offers several advantages:

- Cost effective: Affordable compared to custom aerospace components.

- Rapid development: Extensive community support and SW resources facilitate quick development cycles.

- Versatility: Capable of handling various tasks from data processing to communication.

Studies have focused on understanding the radiation survivability of Raspberry Pi in space, determining its threshold for radiation damage. This research is crucial for ensuring the durability and reliability of CubeSats [9.7]. For example, the development of highly integrated CubeSat platforms, such as ESTCube-2, highlights the advancements in onboard processing for small satellite missions [9.8].

Figure 9.3 The GASPACS.

NASA

9.3.
Integration of Health Management into Spacecraft Design

It is important to ensure comprehensive health monitoring from the mission's inception through ongoing operations. In the early mission analysis phase, the team defines the service

requirements, incorporates them into orbit and trajectory considerations, and identifies critical parameters for health assessment, laying the groundwork. As the overall spacecraft design takes shape, a robust architecture is planned. This involves specifying processors or components in the satellite product tree. It also includes allocating functions to OBCs and bus equipment. Additionally, determining crucial sensors for health monitoring and defining data pools for health variables in the OBSW are part of this process. We can refine the health management details further. This includes allocating functions to OBCs and bus equipment, determining crucial sensors for health monitoring, defining data pools for health variables in the OBSW, finalizing interfaces, integrating commands and telemetry parameters, developing verification and validation methods, and implementing redundancy configurations. A hybrid modular redundancy can offer promising solutions for enhancing reliability in multi-core computing clusters in space environments [9.9].

The architecture needs to accommodate various functions. These include central OBSW, DH SW, health management functions, sensor management, telemetry and telecommand (TMTC), and security functions. The HW and SW choices have a crucial impact on the deployment of any functionalities. The architectural decision involves determining the deployment of health management functions. This can be done by combining different processors or using a single processor with integrated functionalities. For example, the OBSW and DH can be placed in a dedicated processor, with health management functions offloaded in a coprocessor. This setup allows the OBC tasks and system DH to run on a simpler dedicated processor, while other functions operate on a coprocessor with better performance capabilities. The performance requirements for the coprocessor are relatively relaxed because of its dedicated HW, offering a safe solution by isolating critical functions.

The other option involves having a single processing system where everything is integrated. This option risks overloading the processor with all the SW tasks, leading to potential performance issues.

However, it can simplify the architecture, especially if the CPU load is sufficient for both OBC and health tasks. The challenge remains in isolating critical functions to ensure reliable operation. Figure 9.4 shows a few configurations for narrowing down the architectural trade-off at the technology level.

Figure 9.4 System-level architectures.

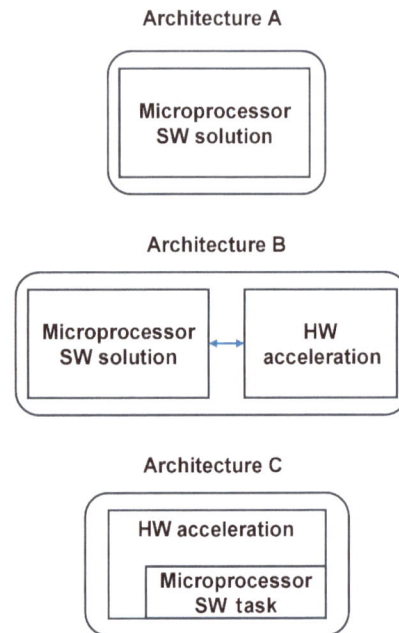

9.3.1.
Architecture A: A Powerful Microprocessor

A single-processor approach for health management during space missions offers a balanced solution between system simplicity and efficiency. However, it requires careful consideration of processor capabilities and resource management to ensure effective health monitoring and prognostics. This architecture offers several advantages and challenges.

Advantages:

- Simplified system complexity: By using a single processor, the system architecture becomes less complex, reducing potential points of failure and simplifying maintenance and updates.

- Energy efficiency: Multiple devices are less energy-efficient than a single processor, which is a crucial factor in space missions where power resources are limited.

- Integrated task management: A unified system manages all health management related tasks, including data collection, analysis, and decision-making, ensuring cohesive and coordinated health management.

Challenges:

- High CPU load: The processor must handle all prognostics and health management (PHM) tasks, potentially operating at near full capacity. This could limit its ability to process extensive diagnostic and prognostic algorithms necessary for effective health management.

- Lower system frequency: Given the current limitations of space-grade processors, the system might operate at a lower frequency compared to architectures with dedicated coprocessors.

- Resource allocation: The processor must manage interfaces with other systems while executing complex algorithms. This multitasking requirement could reduce the resources available for in-depth health analysis and prediction.

9.3.2.
Architecture B: A Powerful Processor and a High-Performance Accelerator

A combined HW/SW solution offers a robust solution for balancing enhanced performance with functional flexibility. AI accelerators have become integral to onboard processing, providing enhanced capabilities for satellite communications and real-time data analysis [9.10]. However, the increased power demand and system complexity require careful consideration to ensure the architecture aligns with mission constraints and objectives.

Advantages:

- Enhanced performance: The HW accelerator offloads computationally intensive tasks, such as image processing or data analytics, from the main processor. This division of labor allows the processor to handle other functions efficiently.

- High-frequency operations: The HW accelerator design enables high-frequency performances, which are crucial for real-time health monitoring and diagnostics in space missions.

- Programmable flexibility: The HW accelerator offers programmable routing, providing flexibility in implementing interfaces, communications, and data arbitration. This adaptability is essential for accommodating various tasks and evolving mission requirements.

- Functional isolation: By distributing tasks between the processor and the HW accelerator, this architecture inherently isolates functions. Segregation minimizes error propagation in the system.

Challenges:

- Increased power consumption: Using both a processor and a HW accelerator results in higher power usage compared to a single-processor solution. Increased power consumption in space missions, where power resources are limited, could pose a significant drawback.

- System complexity: This architecture introduces additional complexity in terms of system integration and synchronization between the processor and the HW accelerator, requiring sophisticated coordination mechanisms.

- Fault tolerance management: Although functional isolation provides benefits, managing fault tolerance across multiple components can be challenging. Ensuring robustness and reliability in the face of potential HW failures or anomalies becomes more complex.

9.3.3.
Architecture C: An Integrated Solution Combining HW Accelerator with Embedded Processors

This is a forward-looking SoC approach that combines the advantages of separate processor and HW accelerator configurations while minimizing their drawbacks. Its potential for high performance, combined with lower power consumption and efficient data management, makes it a promising solution for health management during space missions, pending further development and qualification.

Advantages:

- Optimized performance: Like architecture B, this approach enables high-frequency operations essential for real-time monitoring, offloading intensive tasks to the HW accelerator.

- Reduced power consumption: By integrating the processor and HW accelerator into a single SoC, this architecture consumes less power than having separate devices, a critical factor in space missions.

- Efficient data management: The integration reduces the need for extensive data exchange between separate components, further conserving power and enhancing system efficiency.

- Functional isolation and flexibility: Despite being on a single chip, the architecture can still isolate functions for error minimization and offers flexibility for implementing various interfaces required for tasks.

- Potential for high-performance SoC options: Emerging technologies, such as NanoXplore's NG-ULTRA or Xilinx's radiation-tolerant RFSoC, Zynq, and Versal ACAP, show promising prospects for this architecture.

Challenges:

- Space-grade qualification: To date, there are no space-grade SoCs fully qualified, though some are on the near-future roadmap. This lag in qualification could initially limit the adoption of this architecture in space missions.

- Integration complexity: Designing and manufacturing a SoC that effectively integrates both processor and HW accelerator capabilities, while meeting space-grade standards, presents significant engineering challenges.

- Compatibility and standardization: Integrating this new module with existing avionics architecture components requires careful consideration of compatibility and standardization. Ensuring that these new systems can seamlessly integrate into established frameworks is crucial for their successful deployment in space missions.

Ultimately, the choice of architecture must satisfy the mission requirements in terms of real-time capabilities, operational deadlines, and crucial technical budgets, such as power consumption.

As the spacecraft progresses toward launch, it needs to be verified and validated (Phase D). We must test various functionalities and automated sequences for health monitoring. Upon establishing ground station contact, we start real-time health assessment. Postlaunch, this becomes an integral part of routine operations, with continuous monitoring of spacecraft health and executing anomaly detection and recovery procedures. The team performs preprocessing and analysis of the spacecraft health data for trend analysis, and they update algorithms based on in-flight experiences. Adaptation and optimization can occur continuously throughout the mission. We analyze the health data, incorporate lessons learned into future designs, and optimize algorithms for efficiency. This iterative process ensures that health management remains dynamic, adapting to evolving mission requirements and technological advancements.

9.3.4.
OBSW

OBSW is essential for spacecraft because it detects problems, fixes them, and keeps the mission on track [9.11]. OBSW operates at multiple levels to manage anomaly detection, component isolation, and recovery actions, ensuring spacecraft health and functionality.

- Top OBSW level: This module serves as the overarching health monitor and recovery

action controller for the entire system. It receives information from lower-level modules and decides on recovery actions based on the severity of the issues detected.

- Lower OBSW level: equipment handlers: Monitor data communication between the OBC and spacecraft equipment. They report anomalies, such as equipment failures to respond or mode transition failures.
- Application-specific OBSW (payload, power, thermal, AOCS): Include a failure detection layer to identify equipment performance and logical failures.

These levels may have implications for spacecraft operations, especially in scenarios where recovery actions involve transitioning the spacecraft to Safe Mode or other operational modes. Either way, it requires the generation of status telemetry for ground control to provide visibility into the state of the spacecraft.

Handling SW and HW failures can also be a challenge. HW-detected memory failures, like SEUs, pose challenges in many electronics. Therefore, many modern memory chips feature error detection capabilities. While single-bit failures can be autonomously corrected, OBSW can use checksums to handle double failures. Since modern onboard OSs lack memory virtualization, addressing such issues will require HW reconfiguration. Certain HW-detected failures, such as uncorrectable register file SEU errors, occur directly within the processor. These instances may seem closer to the OBSW but present a challenge to resolve because of potential data corruption before detection. The best achievable outcome involves storing alerts in nonvolatile memory for ground diagnosis and triggering OBC reconfiguration, irrespective of spacecraft operational mode.

9.3.4.1.
SW Considerations

To develop onboard health management SW for a spacecraft, specific FDIR functions need to be incorporated within the OBSW. These functions monitor component health and address potential issues. Functions need to have specific operational modes that are based on mission requirements, the criticality of telemetry data, and the desired balance between continuous monitoring and diagnostic activities. The data variables involved in this process include:

- High-priority telemetry (HPTM) provision to ground:
  - By the OBSW: The spacecraft's OBSW sends important telemetry data directly to the ground.
  - Non-OBC units: The OBC and SW route telemetry from other spacecraft units.
  - HW units: Certain HW components, like the CCSDS processor, independently provide telemetry.
- Housekeeping telemetry packets generation:
  - Structure IDs (SID): Serve as a blueprint that defines what information should be included in telemetry packets.
  - Parameter types: Specifies the nature of parameters (Boolean, integer, real, etc.) in the telemetry.
  - Calibration ID: Provides calibration details for accurate interpretation.

- Cyclic generation of telemetry packets:
  - Internal data pool variables: The OBSW generates periodic telemetry from its internal data.
  - External equipment-provided packets: collects and transmits telemetry data.
- Limit monitoring:
  - OBSW and equipment variables: Constantly checks and monitors predefined limits in both SW and equipment variables.
- Logging mechanisms for events:
  - Event logging: Records important events like system failures or limit breaches.
  - Safeguard memory: Stores these events securely for postanalysis even if the OBC reboots.
  - Reconfiguration log identification: Keeps a log of any reconfigurations made, helping to understand changes after a reboot.
- Equipment health status parameters:
  - Storage in spacecraft configuration vector: Maintains parameters describing the health of various spacecraft components.
  - Ground management: Allows ground control to access critical status information.
  - Packets, logs, and their relevance.

Table 9.5 provides a detailed breakdown of the key elements and considerations using the highlighted variables, offering a comprehensive overview of its functionality and integration with broader spacecraft design principles.

Table 9.5 Key aspects of OBSW development.

Aspect	Description
Definition	FDIR is the spacecraft's healing mechanism, designed to detect, isolate, and recover from failures autonomously.
Hierarchy	A clear hierarchy identifies and manages different failures. It allows for the escalation of failure handling from lower to higher levels and recovery from higher to lower levels. For instance, a system-level failure may trigger reconfigurations on Level 3, such as power failures. Conversely, recovery often involves transitioning the entire spacecraft to Safe Mode, which reconfigures the entire system, including buses and power lines, to the redundant side. This comprehensive approach ensures effective FDIR operations and enhances spacecraft resilience in the face of anomalies.
Autonomous Safe Mode	The spacecraft must autonomously transition to Safe Mode when needed. This will ensure the spacecraft operates with maximum redundancy and minimum resource consumption.
Ground interaction	• Ground control can submit commands, analyze detailed status reports, and review the event history for precise identification of failures. • Adjusting the operational limits can prevent future Safe Mode triggers caused by equipment degradation.
Safe mode characteristics	Safe mode is carefully defined to operate with maximum redundancy and minimal resource consumption. The transition involves clearing HW and SW interfaces, preventing the resumption of interrupted functions during or after the FDIR process. Safe mode will also trigger during specific mission phases that may have constraints, inhibiting deployments or actuator control until certain stages are reached.
Integration with other concepts	These elements closely link to other spacecraft design concepts, including observability and redundancy concepts. These concepts work together to ensure the spacecraft's operational success and adaptability.

9.3.5.
The Importance of "Safe Mode"

The concept of Safe Mode is integral to spacecraft operations, serving as a fail-safe mechanism in response to identified failures. To understand its importance, we need to acknowledge that transitioning the spacecraft into this mode will disrupt all onboard functions. Therefore, the decision to trigger a Safe Mode operation is pivotal, with automation contingent on the swiftness of ground systems in identifying issues and starting isolation and recovery processes. Each subsystem will have configuration guidelines. On the redundant side, the OBC will handle important tasks like managing the mass memory unit's housekeeping and protecting memory for the spacecraft configuration vector, along with the CCSDS processing unit.

This process ensures attitude stability and adequate power generation through the solar arrays. The OBSW will also monitor spacecraft limits with dedicated settings. During Safe Mode, several subsystems, including the main data bus, the OBC I/O unit, and the Power Control and Distribution Unit (PCDU), operate on their redundant side. The PCDU runs minimally on its redundant controller, with the Latching Current Limiter (LCL) bank switching applied only if the PCDU experiences a failure. The LCL protects electrical circuits by limiting current and isolating faulty sections in the event of an overcurrent.

Power bus voltage is carefully monitored to ensure it stays within Safe Mode limits. The AOCS, a critical component, also operates in redundancy during Safe Mode, including the reaction control system. To conserve energy during periods of interrupted mission activity, unnecessary equipment is either powered down or switched to a low-resource state.

The recovery process involves configuring the spacecraft for nominal operations, selecting the appropriate OBSW boot image, rebooting the desired OBC redundancy with the patched or selected OBSW image, and loading the spacecraft control vector (SCV). Once the OBSW applies the SCV and redundancies are set to the correct configurations, the spacecraft transitions through system mode changes to return to nominal operations. This includes switching the AOCS subsystem to normal mode, reconditioning resources, and loading a new mission timeline.

This systematic approach ensures a smooth and controlled recovery from Safe Mode, allowing the spacecraft to resume regular operations efficiently.

9.3.6.
Onboard Implementation

When considering onboard implementation, it is essential to consider a variety of factors, including previous implementations, state-of-the-art processing architectures, and mission-specific constraints. The trade-off analysis for these architectures should consider various quantitative and qualitative parameters. These include development time, cost, scalability, weight, transistor area, power requirements, clock speed, and future missions. This emphasizes integrating different HW

components to enhance performance, reliability, and functional mapping. As discussed earlier, various architectures incorporate generic HW accelerators. These present different solution implementation options:

- FPGA-based architecture: Offers flexibility and reconfigurability, suitable for real-time processing and fault-tolerant designs.
- GPU-based architecture: Ideal for high-throughput data processing, particularly effective in image and signal processing tasks relevant to health management.
- Multicore DSP-based architecture: Efficient for parallel processing, useful in tasks requiring simultaneous data analysis from multiple sensors or subsystems.

Each of these has unique advantages and trade-offs, leading to a careful evaluation based on the specific needs and constraints of the space mission. This holistic approach ensures that the chosen architecture not only meets the current requirements but is also viable and effective for the long-term goals and evolution of space missions. Table 9.6 summarizes the relevance to spacecraft health management.

Each architecture has implications for spacecraft health management systems to process capabilities, power consumption, and suitability for specific tasks. The decision on which architecture to use will depend on the specific requirements of the system, including the need for real-time data processing, system integration constraints, and overall mission objectives.

Table 9.6 Characteristics of various architectures for health management.

Architecture type	Key characteristics	Limitations	Suitability for monitoring
Multiple processors	SW parallelization, communication through message passing/shared memory.	• Performance affected by synchronization delays and inefficiencies in the general-purpose programming model. • Space-grade processors typically show lower performance than commercial variants.	Suitable for parallel processing; may have speed/efficiency limits.
Processor + multiple DSPs	Enhanced processing with DSP cores for efficient signal processing.	• Performance issues can arise at higher sample rates or with multichannel requirements.	Effective for specific signal processing tasks.
FPGA	High parallelization, various configurations (single, multi-FPGA, etc.).	• Finite resources. • Less efficient for sequential tasks compared to parallel processing tasks. • Do not optimally support the floating-point operations.	Ideal for intensive data processing tasks.
SoC	Integrated CPU with FPGA, DSP, or GPU cores for balanced performance.	• Integrating multiple functionalities (processor, FPGA, DSP, GPU) in a single chip can be complex and challenging. • Lacks the flexibility of separate components in terms of upgradability and customization. • Availability of space-qualified SoC variants is limited. • The development and qualification process can be lengthy and expensive.	Integrated approach for complex tasks.
ASIC	Custom HW design, high speed, and low power consumption include ARM-based options.	• Involve significant upfront costs, making them less viable for low-volume applications. • Resolving design errors or a requirement change can be costly. • The development cycle is typically longer than for programmable devices.	High performance for tasks but with higher development costs.

The system requires a thorough verification process to guarantee reliable performance, identify issues, and maintain spacecraft health and safety throughout its mission. This includes a range of testing methodologies—from unit and integration tests to system and acceptance tests—to ensure:

- Assessing the balance between HW acceleration and SW control.
- Dictating the level of verification and quality assurance required, for example, as defined by the European Cooperation for Space Standardization (ECSS) to ensure quality and consistency.
- The evaluation of processor-plus-coprocessor architectures, determining coding languages, and choosing to operate systems (usually RTOSs).
- The HW/SW meets all requirements and functions correctly.
- We execute and test every line of code, particularly after modifications, to maintain the integrity of the SW.
- Comprehensive testing of all customer requirements and acceptance tests.

While this list is nonexhaustive, it captures the elements that will ensure that the spacecraft's health management system is robust, reliable, and capable of operating effectively during missions. The idea is to progressively move from the system modeling environment to the real-world implementation. Table 9.7 presents a summary of quick reference.

Table 9.7 Key aspects of OBSW implementation.

Aspect	Description
SW/HW integration	Integrating PHM functions with spacecraft infrastructure
Partitioning	Balancing HW acceleration with SW control
Standards adherence	Following ECSS or equivalent standards
Criticality evaluation	Assessing the impact of potential failures
Design decisions	Selecting HW, coding languages, OS
Real-time system feasibility	Adapting prototype to real-time requirements
SW implementation	Managing sensor data, interfaces, FDIR
Prototyping and profiling	Assessing performance on selected HW
Schedulability analysis	Timing and synchronization of SW tasks
V&V campaigns	Comprehensive testing methodologies
Code coverage	Ensuring all code lines are tested
System tests	Final testing phase including customer acceptance

9.4.
Ongoing Challenges with AI

The growing demand for DL in commercial applications has led many electronics manufacturers to invest in specialized HW accelerators. Specialized HW accelerators enable electronics manufacturers to execute DNNs efficiently at the edge, offering impressive trade-offs in performance and power consumption at a reduced cost. This makes them capable of outperforming standard graphics processing units and CPUs in both edge and cloud-based ML. To highlight the potential of such devices for space applications, we can analyze their performance using a metric that considers megapixels per second per watt per kilogram per dollar (Mpixels/s/W/kg/$). This metric is relevant for several reasons:

- Computing efficiency: The Mpixels/s/W aspect indicates computing efficiency, crucial for applications like SmallSats where power efficiency is vital.

- Mass reduction: Minimizing the weight of computing platforms is essential due to its direct impact on launch costs and the potential for miniaturization.

- Cost reduction: Lowering costs is critical in the New Space era, as payload command and data processing platforms can significantly influence the total cost of a satellite.

For example, the ESA's upcoming lunar regolith miner mission's computing platform, which includes a computer vision subsystem, shows a stark contrast when compared to commercial solutions. The estimated cost of this subsystem is over $200,000. The system processes less than 4 FPS at a 1 Mpixel frame resolution, weighs approximately 35 kg, and dissipates around 60 W. In comparison, the commercial Myriad 2 SoC achieves a performance metric of 250,000, highlighting the advantages of a lightweight and affordable COTS solution for space applications. Its design specifically focuses on achieving processing efficiency per watt and per dollar. ESA has a history of using commercial parts in spacecraft, including critical applications, achieved through careful selection, qualification,

© SAE International.

and screening. However, to extend the applicability of DL in missions with more stringent radiation resistance requirements, alternative approaches are being explored. For instance, Blacker et al. developed a tool to facilitate the porting of DNNs onto the LEON 3 processor, a space-qualified device. This shows a growing interest in integrating advanced DL capabilities into more radiation-resistant, space-grade HW.

9.4.1.
Toward Edge Computing

Currently, most AI applications in space limit themselves to processing data offline rather than directly on spacecraft. This is mainly because it is difficult to adapt DL networks to older spacecraft HW, which often lacks the performance needed for even basic tasks. For instance, the memory requirements of accurate models often exceed what satellites can handle. Running these models requires a lot of computational power, which is a problem in space because of power constraints and the difficulty of dissipating heat.

Dedicated AI platforms like Myriad 2 have appeared to address these issues. They are designed to balance model complexity, processing speed, and power consumption efficiently while running complex CNNs. Strategies like model compression, including knowledge distillation, quantization, and pruning, can reduce model size without losing accuracy. Another challenge is the unpredictability of ML algorithms, which is a concern in space missions where safety is paramount.

To minimize risks, many applications could limit themselves to less critical tasks, such as object detection on sensor data. For less critical missions, fault-tolerant systems could supervise the use of COTS processors and FPGAs. Training deep networks is also challenging, especially

when suitable training datasets for new equipment are unavailable. Usually, researchers train models on Earth using advanced HW, which raises questions about the suitability of these models for space missions. One solution is the ability to reconfigure models during a mission using modern COTS ASICs, which would allow for updates and adaptations in orbit. Introducing such technology would indeed revolutionize data processing. Rather than focusing on compressing and storing raw images for transmission to Earth, an edge device could preprocess images in space, generating metadata and enabling selective data transmission. This approach could significantly reduce bandwidth requirements, especially for images that are largely covered by clouds.

An AI-based architecture could also benefit applications like "cloud detection." The HyperScout 2 satellite, part of the PhiSat initiative, is expected to implement this concept. It will feature CloudScout, the first in-orbit demonstrator using DNNs for processing hyperspectral images. CloudScout, a CNN-based algorithm, will identify cloudy and not-cloudy images, allowing the satellite to discard images covered by clouds, saving data. The CloudScout DNN will process its inference through the Myriad 2 accelerator, which was chosen for its proven radiation resistance. This aspect has been validated through specific de-risking test campaigns. Besides cloud detection, onboard AI can also be beneficial in fault detection and management. For example, a method based on CNNs can detect faults in satellite images and determine whether the level of corruption affects their usability for end users. Similar to cloud detection, generating metadata for these images can lead to discarding unusable images, thus saving data.

The edge computing approach improves respon-siveness, which is another advantage of onboard processing. Knowing the content of images in advance allows for prioritizing downloads based on their importance. This is crucial for time-sensitive applications like fire or oil spill detection, where quicker data transmission can significantly reduce disaster response times. The AI-based architecture holds potential for various other EO applications, such as surface terrain classification, anomaly detection, change detection, and object detection.

Preprocessing images on board, while not always immediately helpful, can be a valuable first step in extracting metadata and identifying images of interest for content-based downloading. This approach can help manage capacity effectively. For instance, in classification algorithms, metadata could simply be the classification result. For tasks like image segmentation, metadata might summarize the image content, such as the percentage of different land parti-tions or the dominant land type.

SoCs are known for their high efficiency in AI inference with pretrained networks, particularly in computer vision tasks such as visual odometry, optical flow, mesh-based image warping, stereo pair matching, and key point tracking. Radiation tolerance testing is essential for space HW because of the abundance of charged particles in space that can cause SEEs in silicon. These effects include bit flips (SEUs), latchups (SELs), and functional interrupts (SEFIs). Prolonged exposure can also lead to failures.

However, accurate assessment of a device's performance in space generally requires actual radiation testing in relevant environments. SoCs, such as Myriad 2, present unique challenges in radiation characterization. This is due to their complex architecture, which includes processors, memories, peripherals, mixed-signal elements, and, in the case of FPGAs, look-up tables and reconfigurable blocks. This complexity necessi-tates a variety of testing techniques, especially for multiprocessor SoCs which have more intricate data paths and shared components. Research under a Large Ion Collider Experiment (ALICE) project at the European Organization for Nuclear Research (CERN), for example, has focused on radiation effects on FPGA-based systems.

This includes testing Xilinx Virtex-II FPGAs used in collider sensing equipment and the Zynq 7000 SoC at 28 nm, with heavy ion testing providing insights into SEE behavior.

Another significant challenge in testing COTS devices for SEL is the number of power rails to monitor and the varying current expectations on each rail. These complexities notwithstanding, the Myriad 2's architecture and memory mapping capabilities make it a promising candidate for robust solutions in space applications.

9.5. Summary

This chapter outlined the challenges emerging from the evolving requirements of the space industry, like coordinating multiple satellites and dealing with limited resources. Onboard processing is an effective solution that has proven successful in space missions. It involves handling data directly on the satellite, reducing the need to transmit large amounts of data back to Earth, and lessening the workload for satellite operators. The increasing number of satellites being launched necessitates the development of new methods to manage and monitor their health through ground stations. AI can help with these tasks because of its ability to handle complex problems by making processes more efficient, like quickly responding to satellite faults or reconfiguration requirements. It can even make certain tasks more accurate, such as predicting problems with the satellite before they happen by analyzing data from its sensors. It can also help in managing EO missions more effectively, like providing quick warnings based on the data it collects.

The trend of using SoCs in aerospace applications is gaining momentum due to several compelling benefits. Mass production for consumer markets makes these components more affordable than custom-built aerospace solutions, providing significant cost efficiency. Their extensive use in various applications ensures a high level of reliability. The rapid development process is facilitated by the availability of extensive documentation, community support, and SW libraries. Additionally, SoCs offer a good balance of processing power and energy efficiency, crucial for space missions with limited power resources. Continuous improvements in ARM technology, combined with the growing demand for low-cost and reliable components, suggest that more aerospace projects will adopt SoCs in the future.

References

9.1. Katz, R.B. and Day, J.H., "FPGAs in Space Environment and Design Techniques," Bibliogov, 2001.

9.2. Mousist, A.D., "Autonomous Payload Thermal Control," arXiv preprint arXiv:2307.15438, 2023.

9.3. Habinc, S., "Suitability of Reprogrammable FPGAs in Space Applications," Feasibility Report FPGA-002-01, 2002.

9.4. Posso, J., Bois, G., and Savaria, Y., "Mobile-URSONet: An Embeddable Neural Network for Onboard Spacecraft Pose Estimation," arXiv preprint arXiv:2205.02065, 2022.

9.5. Delaune, J., Brockers, R., Bayard, D., Dor, H. et al., "Extended Navigation Capabilities for a Future Mars Science Helicopter Concept," in *2022 IEEE Aerospace Conference (AERO)*, Big Sky, MT, 2020.

9.6. Towfic, Z.J., Ogbe, D., Sauvageau, J., Sheldon, D. et al., "Benchmarking and Testing of Qualcomm Snapdragon System-on-Chip for JPL Space Applications and Missions," in *2022 IEEE Aerospace Conference (AERO)*, Big Sky, MT, 2022.

9.7. Decena, J.C., Wood, B., Martineau, R., Dennison, J. et al., "Detection and Classification of High Energy Beta Radiation Induced Damage of Raspberry Pi Zero Intended for OPAL CubeSat," USU GASPACS, 2018.

9.8. Dalbins, J., Allaje, K., Iakubivskyi, I., Kivastik, J. et al., "ESTCube-2: The Experience of Developing a Highly Integrated CubeSat Platform," in *2022 IEEE Aerospace Conference (AERO)*, Big Sky, MT, 2022.

9.9. Rogenmoser, M., Tortorella, Y., Rossi, D., Conti, F. et al., "Hybrid Modular Redundancy: Exploring Modular Redundancy Approaches in RISC-V Multi-Core Computing Clusters for Reliable Processing in Space," arXiv preprint arXiv:2303.08706, 2023.

9.10. Ortiz, F., Monzon Baeza, V., Garces-Socarras, L.M., Vásquez-Peralvo, J.A. et al., "Onboard Processing in Satellite Communications Using AI Accelerators," *Aerospace* 10, no. 2 (2023): 101.

9.11. Gómez, P., Östman, J., Shreenath, V.M., and Meoni, G., "PAseos Simulates the Environment for Operating multiple Spacecraft," arXiv preprint arXiv:2302.02659, 2023.

This section highlights specific technologies, practices, and challenges unique to each category. This includes protecting against both external cyber threats and internal system failures that could compromise mission success. The growing risk of cyberattacks is attributed to factors such as increasing interconnectivity, technological advancements, and the construction of space stations and infrastructures before cybersecurity became a global concern, as highlighted in a recent study [10.1]. As a result, while these systems are sophisticated, they are also quite susceptible to cyberattacks. With the increase in private service offerings, there are accompanying challenges of cybersecurity (   ).

Table 10.1 Reported challenges in cybersecurity for the space domain.

Service offerings	Challenges	Connections	Unique space mission aspects
Communication security	• Satellite communication encryption. • Antijamming and antispoofing techniques. • Secure telemetry and command protocols.	• Encryption is fundamental for all data transmission. • Antijamming/antispoofing techniques are crucial because of the vulnerability of satellite communications to interference.	• Long-distance communication requires robust encryption because of the increased risk of interception. • Space missions often rely on narrow communication windows, making secure and uninterrupted links vital.
Onboard system security	• Intrusion detection systems (IDSs). • HW-based security modules. • Fault tolerance and redundancy.	• IDS must be lightweight and efficient because of limited onboard computing resources. • HW security modules (HSMs) provide a secure environment for critical operations.	• Spacecraft systems require specialized IDS tailored for space environment constraints. • Redundancy is critical because of the inability to physically repair systems in space.
Data security	• Encrypted data storage. • Access control mechanisms. • Secure data. • Transmission protocols.	• Data encryption protects sensitive mission data. • Access control is essential to prevent unauthorized system manipulation.	• Data storage and transmission face unique challenges because of the space environment, such as radiation effects on storage media.
SW security	• Secure boot processes. • SW update authentication. • Anomaly detection algorithms.	• Secure boot ensures integrity at system startup. • SW updates must be authenticated to prevent malicious code upload.	• SW updates in space must be securely transmitted over long distances and authenticated rigorously because of the inability to manually rectify issues.
Physical system security	• Tamper-proof HW design. • Radiation-hardened components. • Secure HW interfaces.	• Tamper-proof designs prevent unauthorized physical access. • Radiation-hardened components ensure functionality in harsh space conditions.	• Physical security measures must account for the extreme conditions of space, including vacuum, temperature extremes, and cosmic radiation.
Operational protocols	• Mission operation security procedures. • Emergency cybersecurity protocols. • Crew and ground staff cybersecurity training.	• Security procedures guide safe mission operations. • Emergency protocols are essential for quick response to cyber incidents.	• Space missions require specific operational security protocols because of isolation and limited physical intervention capabilities.

10.1.1.
Communication Security

Here, I would like to discuss three aspects:

- Advanced Encryption Standard (AES): AES encryption is integrated into the GPS III satellites, developed by Lockheed Martin for the USAF, to bolster the security of military signals in satellite communications. This encryption ensures that GPS communications are protected against interception and unauthorized access, which is crucial given the widespread use of GPS for both military and civilian applications.

- Antijamming and antispoofing techniques: The GPS Block IIF satellites, which are part of the US GPS, include advanced antijamming capabilities. These satellites use a technique called the M-Code (Military Code), designed to be more resistant to jamming and spoofing attempts. The M-Code improves the security and reliability of military GPS signals, ensuring that they remain accessible even in contested environments.

- Secure telemetry and command protocols: The Mars Reconnaissance Orbiter (MRO), a NASA spacecraft orbiting Mars, uses secure communication protocols for telemetry and command operations. The communication system of MRO includes encryption and authentication mechanisms to protect the data transmitted between the orbiter and Earth. This is vital for the integrity of the mission, as it involves sending scientific data back to Earth and receiving operational commands.

It should be noted that no publicly disclosed instances of major cybersecurity breaches in spacecraft (health management) have led to a complete mission failure or takeover of a spacecraft. However, there have been incidents and concerns raised that highlight the potential risks:

- Tiangong-1 concerns (2011): In 2011, the US–China Economic and Security Review Commission reported to the US Congress that two US government satellites had experienced interference in 2007 and 2008. The report suggested these incidents were consistent with a possible Chinese antisatellite test. However, there was no conclusive evidence, and the satellites' operations remained intact.

- NASA's JPL had a cybersecurity breach in 2018: The breach involved an unauthorized Raspberry Pi computer that was used as a gateway to access JPL's systems. This incident did not directly affect spacecraft but highlighted vulnerabilities in the network that might impact critical systems.

- General concerns in satellite communications: There have been various reports and concerns about the vulnerability of satellite communications to jamming and spoofing. While these have primarily been theoretical or shown in controlled environments, they underscore the potential risks to spacecraft communication systems.

- ISS Malware (2008): The ISS reportedly had a malware infection on laptops intended for routine operations. The malware, known as W32.Gammima, did not specifically target the ISS and did not affect mission-critical systems, but it raised concerns about the need for robust cybersecurity measures, even in space environments.

10.1.2.
Onboard System Security

It focuses on safeguarding the spacecraft's internal systems against cyber threats and ensuring operational continuity. This domain encompasses several key concepts, each playing a vital role in maintaining the integrity and functionality of spacecraft systems. Because of the limited computing resources available in space environments, spacecraft tailor IDSs to be lightweight. These systems often employ anomaly-based detection techniques, which monitor the spacecraft's systems for unusual activities or deviations from normal operational patterns. This approach is effective in identifying potential cyber threats or system malfunctions that could compromise the mission.

HSMs are another crucial aspect of onboard system security. Integration of these modules into the spacecraft's HW provides a secure environment for critical cryptographic operations. HSMs protect sensitive cryptographic keys and ensure secure execution of encryption and decryption processes, which are essential for safeguarding communication and data storage in space missions.

Redundancy is a fundamental principle in spacecraft design, particularly for critical systems such as navigation, communication, and life support. By having redundant systems, spacecraft continue functioning even if one component fails. This redundancy is crucial for mission continuity, as it provides a backup in case of system failures, whether because of natural space hazards, technical malfunctions, or targeted cyberattacks. Examples include:

- The Mars rover missions by NASA, where redundancy in navigation and communication systems has been a key design feature.

This redundancy ensures that the rovers can continue their scientific missions even if one system component fails.

- Using HSMs in commercial satellites, such as those operated by companies such as SpaceX and Blue Origin, to secure communication links and protect onboard data.
- Anomaly-based IDS implementations in the ISS, which monitor the station's network for unusual activities, protecting against potential cyber threats.

The absence of these security measures could lead to severe consequences. Without effective IDS, cyber threats could go undetected, leading to compromised missions or even loss of control over the spacecraft. A lack of HSMs could cause the exposure of cryptographic keys, making encrypted communications vulnerable to interception. Without redundancy, a single system failure could lead to mission failure or, with manned missions, pose a direct threat to astronaut safety.

The potential risks underscore the importance of robust onboard system security. The continuous evolution of cyber threats makes it important for space missions to incorporate advanced security measures to protect against both known and emerging vulnerabilities.

10.1.3.
Data Security

It ensures data confidentiality, integrity, and availability for the spacecraft. This domain encompasses several key concepts, each playing a vital role in safeguarding data against unauthorized access and cyber threats.

Using full-disk encryption is a standard practice to protect data stored on spacecraft. This method encrypts the entire storage medium, ensuring that all data are secured against unauthorized access. Scientific satellites, like those used in EO or deep space exploration missions, utilize full-disk encryption to safeguard sensitive data such as valuable and confidential data payloads.

Organizations implement access control mechanisms, particularly role-based access control (RBAC) models, to restrict access to spacecraft systems and data. RBAC allows for the assignment of permissions based on roles within the mission control team, ensuring that individuals can only access the information and systems necessary for their specific roles. This is crucial in large missions, such as the ISS, where multiple teams and individuals need differentiated access to various systems and datasets.

Remote spacecraft operations utilize secure data transmission protocols, such as Secure Shell (SSH). SSH provides a secure channel over an unsecured network, ensuring the integrity and confidentiality of data transmitted between the spacecraft and ground control. For example, in rover missions on Mars, such as NASA's Perseverance rover, the use of secure protocols is essential to transmit commands and receive data. This ensures that the rover operates as intended and protects the data it sends back from interception or tampering.

If the organization does not adequately implement these data security measures, the consequences could be significant. Inadequate implementation of data security measures could lead to interception and exploitation of sensitive data, compromising scientific integrity, mission objectives, and national security involving military satellites. Inadequate access control could cause unauthorized access to spacecraft systems, potentially leading to malicious commands or data manipulation. Insecure data

transmission could expose communications to interception and tampering, compromising mission control and data integrity. The continuous advancement of cyber threats makes it imperative for space missions to incorporate advanced security measures to protect against both known and emerging vulnerabilities.

10.2.
SW Security

It ensures the integrity and reliability of the SW running on spacecraft systems. This domain encompasses several key concepts, each playing a vital role in safeguarding the spacecraft against SW-related cyber threats and ensuring the smooth operation of its mission [10.2]. The implementation of secure boot processes verifies the integrity of SW on spacecraft startup. This process guarantees the integrity of the spacecraft's SW, ensuring safe and reliable operation. Mars rovers, such as NASA's Curiosity, employ secure boot processes to start up in a known and trusted SW state, even after enduring harsh space conditions.

SW update authentication is another critical aspect, where cryptographic techniques are used to ensure the authenticity of SW updates sent to the spacecraft. This is important for long-duration missions, where SW updates may be necessary to improve functionality or address issues. Cryptographic signatures ensure that updates are trusted and unaltered during transit. The Juno spacecraft, orbiting Jupiter, is an example where authenticated SW updates are crucial because of the extended duration of its mission and the need for periodic SW enhancements.

Anomaly detection algorithms are used to monitor SW behavior for signs of malicious activity or unexpected anomalies. These algorithms can detect unusual patterns that may show a cyber threat or an SW malfunction. The ISS uses such systems to monitor its OBSW. These systems ensure that any anomalies are quickly identified and addressed to maintain the safety and integrity of the station and its crew.

Severe consequences could arise if these SW security measures were absent [10.3]. Without secure boot processes, spacecraft could start with compromised or malfunctioning SW, leading to mission failure or loss of control. Inadequate SW update authentication could allow malicious SW updates to be installed, potentially hijacking or damaging the spacecraft. Lack of effective anomaly detection could cause delayed recognition of SW issues, jeopardizing the mission and the safety of any crew aboard.

10.2.1.
Physical System Security

It focuses on protecting the spacecraft's HW from physical tampering, environmental hazards, and unauthorized access. This domain encompasses several key concepts, each playing a vital role in ensuring the spacecraft's physical integrity and operational reliability.

Tamper-proof HW design is a fundamental aspect of spacecraft security. This involves incorporating physical locks and seals on critical components to prevent unauthorized access or tampering. For instance, satellites used for military or intelligence purposes often feature tamper-proof designs to protect sensitive equipment and data. Reconnaissance satellites provide an example of tamper-proof measures,

as they must prevent adversaries from gaining access to sensitive imaging or signal interception equipment.

Radiation-hardened components are essential in spacecraft design, given the harsh space environment characterized by high levels of cosmic radiation. These components undergo special design to withstand the effects of radiation, which can otherwise cause malfunctions or degrade the performance of electronic systems. The James Webb Space Telescope (JWST), for example, uses radiation-hardened components to ensure its sensitive instruments can operate reliably in the radiation-rich environment of space.

Secure HW interfaces are crucial for maintaining the security of spacecraft systems. These interfaces prevent unauthorized devices from connecting to the spacecraft's systems, which might introduce vulnerabilities or malicious SW. The ISS employs secure HW interfaces for all external connections, ensuring that any equipment or modules added to the station meet strict security standards.

If we do not adequately implement these physical system security measures, the consequences could be significant. Without tamper-proof designs, critical spacecraft components could be vulnerable to unauthorized access or sabotage. Without radiation-hardened components, spacecraft systems could fail because of the harsh space environment, potentially leading to mission failure.

Insecure HW interfaces could allow unauthorized access to spacecraft systems, posing a risk to the mission and, with crewed spacecraft, to astronaut safety.

10.2.2.
Operational Protocols

These ensure the overall cybersecurity and safety of space missions. These protocols encompass various practices and procedures that organizations use to maintain secure operations, respond to emergencies, and ensure personnel are trained to handle cybersecurity threats. Mission operation security procedures are a cornerstone of these protocols.

They include comprehensive guidelines for secure communication and DH during missions. For instance, in the Mars rover missions, NASA implements strict security procedures for communicating with the rovers and handling the scientific data they transmit back to Earth. These procedures ensure that NASA encrypts communications and maintains data integrity, protecting against potential cyber threats or data corruption.

Emergency cybersecurity protocols are critical for responding to incidents in space. The protocols detail steps for handling cybersecurity breaches, like unauthorized access or malware attacks. The ISS operations provide an example of implementing emergency protocols to handle cybersecurity incidents. This action is taken to protect the station's systems and ensure the crew's safety. These protocols include isolating affected systems, assessing the impact of the breach, and implementing measures to restore secure operations.

Training for crew and ground staff is essential in recognizing and mitigating cyber threats.

This training ensures that all personnel involved in the mission are aware of potential cybersecurity risks and know how to respond appropriately. As an example, astronauts on the ISS receive training on how to stay safe online, like identifying suspicious activities and keeping sensitive information secure. Similarly, mission control centers train ground staff in cybersecurity awareness to ensure they can support secure mission operations and respond effectively to any incidents.

Inadequate implementation of these protocols could have severe consequences. Inadequate security procedures could lead to data breaches, compromising mission integrity and potentially leaking sensitive information. Poorly defined emergency protocols could result in ineffective responses to cybersecurity incidents, exacerbating their impact. Lack of proper training could leave crew and staff ill-prepared to recognize or respond to cyber threats, increasing the risk of successful cyberattacks.

10.3.
The Role of Resource Management

This involves organizing various aspects of managing resources, playing a vital role in the overall health and functionality of the spacecraft. Table 10.2 categorizes these concepts.

Table 10.2 Reported challenges in resource management for the space domain.

Category	Challenges	Connections	Not connected aspects
Power management	• Solar panel efficiency. • Battery storage and management. • Power allocation and distribution.	• Solar panel efficiency directly affects battery storage capacity. • Power allocation depends on battery management for optimal distribution.	• Power management is independent of other resource management systems like DH.
Thermal control	• Heat shield technology thermal regulation systems. • Radiator panels.	• Thermal regulation systems often work in tandem with radiator panels to manage spacecraft temperature. • Heat shield technology is crucial during reentry phases and less connected to regular thermal control.	• Thermal control has minimal direct interaction with crew resource management.
Data management	• Data storage solutions. • Data transmission efficiency. • Onboard data processing.	• Data storage capacity affects onboard data processing capabilities. • Data transmission efficiency is crucial for sending processed data back to Earth.	• Data management is typically separate from power management but may require power allocation for data processing tasks.
Crew resource management	• Life support systems. • Crew health monitoring. • Spacecraft habitability.	• Crew health monitoring and life support systems are closely connected. • Spacecraft habitability directly affects crew health and performance.	• Crew resource management is largely independent of thermal control, though environmental comfort is a minor connection.

© SAE International

10.3.1.
Power Management

In space, the spacecraft needs to generate, store, and distribute power to various systems on its own. In Figure 10.1, solar panels capture solar energy and convert it to electrical power. The power control unit (PCU) regulates this power, ensuring safe levels. Batteries store excess power for use when solar energy is not available. The distribution system allocates power to various spacecraft subsystems. Subsystem power regulators ensure each subsystem receives the correct power. The thermal control system manages heat from power generation and storage. The monitoring and control system oversees the entire power management process, ensuring efficiency and safety.

The primary source of power for many spacecraft, especially those in Earth orbit or on interplanetary missions, is solar panels. These panels convert sunlight into electrical energy. For example, the ISS uses large solar arrays to generate power. The efficiency of these solar panels is crucial, as it determines how much power the spacecraft can generate.

Battery storage and management are also key components. Batteries store energy for spacecraft when not in direct sunlight, like during orbit on a planet's night side. Efficient battery management ensures that there is always enough power for critical systems, even when solar energy is not available. Mars rovers, like Curiosity and Perseverance, use battery power to operate at night and during dust storms when solar power is reduced.

Figure 10.1 Various subsystems/modules in a spacecraft.

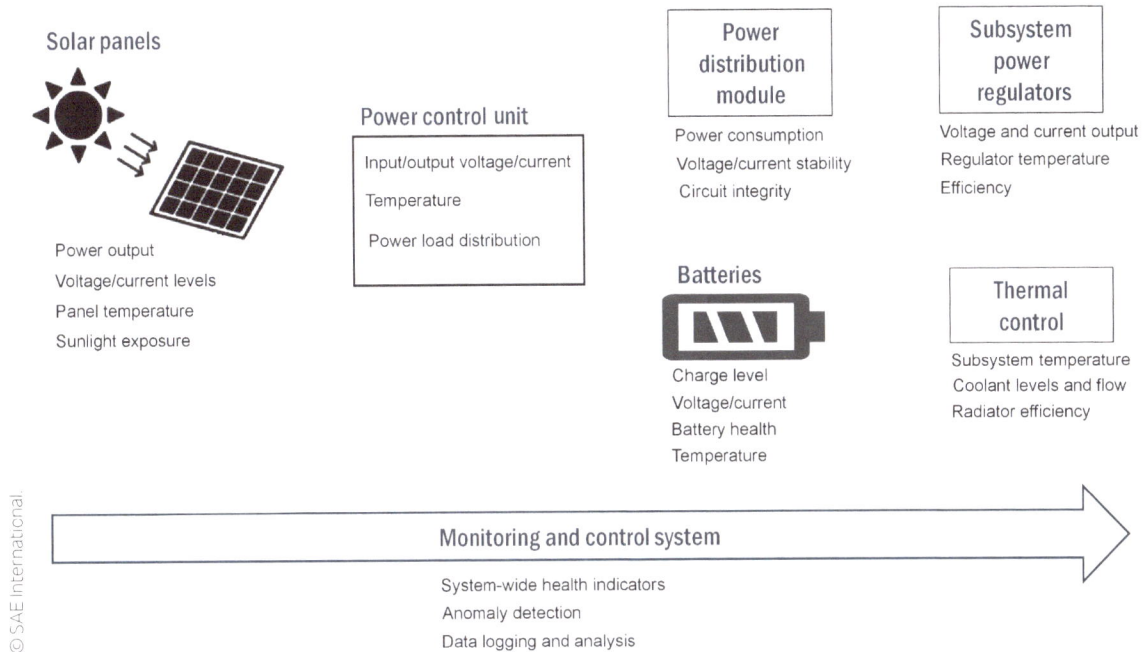

Solar panels

Power output
Voltage/current levels
Panel temperature
Sunlight exposure

Power control unit

Input/output voltage/current

Temperature

Power load distribution

Power distribution module

Power consumption
Voltage/current stability
Circuit integrity

Subsystem power regulators

Voltage and current output
Regulator temperature
Efficiency

Batteries

Charge level
Voltage/current
Battery health
Temperature

Thermal control

Subsystem temperature
Coolant levels and flow
Radiator efficiency

Monitoring and control system

System-wide health indicators
Anomaly detection
Data logging and analysis

Power allocation and distribution systems efficiently distribute power to various spacecraft systems as needed. This involves prioritizing power distribution to critical systems, such as life support in crewed spacecraft or scientific instruments in unmanned missions. The HST, for example, uses a power management system to allocate power to its various scientific instruments and onboard systems.

If power management systems fail, the consequences can be severe. A spacecraft could lose its ability to communicate, its scientific instruments could become inoperable, or, with crewed spacecraft, life support systems could fail. Sometimes, power management issues have affected space missions. For example, the Mars rover Opportunity lost power during a massive dust storm on Mars in 2018, leading to the end of its mission. Similarly, power limitations forced the ESA's Rosetta spacecraft to enter hibernation mode during its mission as it moved away from the Sun.

10.3.2.
Thermal Control

The thermal control manages the extreme temperature variations in space, which can range from intensely hot when exposed to the sun, to extreme cold in the absence of sunlight. Monitoring key parameters is crucial. Some of the notable technologies include:

- Heat shields: To monitor surface temperature and heat flux to ensure protection against extreme heat during reentry.

- Thermal regulation systems:

  - To check temperature control and power consumption of active elements (heaters/coolers).

- To insulate the integrity and thermal resistance of passive elements.

- Radiator panels: To monitor radiating efficiency and surface temperature to ensure effective heat dissipation.

- Temperature sensors: To provide data on ambient temperature around the spacecraft and the temperature of specific components.

- Control unit: To monitor the health of the control system and its response time to temperature changes.

- Coolant loops/heaters: To monitor coolant temperature and flow rate, and heater power levels to maintain operational temperatures.

It typically includes components like heat shields, thermal regulation systems, and radiator panels. Heat shields play a crucial role in protecting crewed spacecraft, like NASA's Space Shuttle or SpaceX's Crew Dragon, during reentry. They safeguard the spacecraft from the extreme heat produced by atmospheric reentry. Thermal regulation systems, which can include active elements like heaters and coolers and passive elements like thermal blankets, maintain a stable temperature inside the spacecraft. Radiator panels are used to dissipate excess heat into space, a critical function for spacecraft like the ISS, where electronic systems and human occupants generate significant heat.

Monitoring subsystem temperatures, coolant levels, and radiator efficiency is vital. For instance, the Mars rover Curiosity uses a thermal control system to manage extreme temperature variations on Mars, ensuring its instruments and systems function correctly.

If thermal control systems fail, the consequences can be severe. Overheating can lead to system malfunctions or damage, while extreme cold can cause systems to freeze and stop working. In crewed spacecraft, failure in thermal control systems can be life-threatening.

Sometimes thermal control issues have affected space missions. For example, the Apollo 13 mission faced critical thermal control challenges after an oxygen tank explosion. The crew had to use the LM creatively as a "lifeboat," carefully managing its thermal control to maintain livable temperatures. Another example is the Galileo spacecraft en route to Jupiter, which experienced problems with its main antenna. One contributing factor was the thermal environment it encountered, which was colder than expected and affected the deployment mechanism.

10.3.3.
Data Management
This encompasses the collection, storage, processing, and transmission of data generated by the spacecraft:

- The data storage solutions perform various tasks: They monitor storage capacity utilization to ensure enough space for new data, conduct data integrity checks to prevent corruption, and monitor read/write speeds for efficiency.

- Data transmission protocols involve several key tasks: checking transmission bandwidth to ensure efficient data transfer, monitoring signal integrity for reliable communication, and tracking communication latency, which is particularly crucial for deep-space missions.

- Onboard data processing: Monitor processor performance to ensure timely data processing, check data processing efficiency to optimize computational tasks, and track the error rate in computation to maintain data accuracy.

The spacecraft can include some onboard data storage solutions, efficient data transmission protocols, and advanced onboard data processing capabilities. For example, the Mars Science Laboratory rover, Curiosity, uses sophisticated data management systems. It collects scientific and operational data, stores them onboard, and transmits them back to Earth using a combination of direct communication with the DSN and relays through orbiters. The rover's OBC processes these data to execute commands and navigate the Martian terrain.

Efficient data transmission is necessary, especially for deep-space missions where communication delays can be significant. The New Horizons spacecraft, which flew by Pluto, utilizes a data transmission protocol that is optimized for long-distance communication. This protocol guarantees valuable scientific data transmission despite vast distances. If data management systems fail, the consequences are significant. Loss of data can mean missing out on key scientific discoveries or losing track of the spacecraft's health and status. Inefficient data processing can lead to delays in mission operations, while failures in data transmission can cause a complete loss of communication with the spacecraft.

Sometimes data management issues have affected space missions. For example, the HST experienced a problem with its data formatting HW in 2008, which temporarily halted data transmission to Earth. Another example is the Galileo spacecraft en route to Jupiter, which faced challenges with its HGA, affecting its data transmission capabilities.

10.3.4.
Health Data Flow Design

In spacecraft health management, data flows seamlessly from individual components to subsystems and then to the overarching system, later facilitating a comprehensive approach to monitoring and diagnosis. This process involves generating and processing health data at each level, sharing these data for collective analysis, and organizing them efficiently to ensure the spacecraft's optimal performance and safety. Figure 10.2 illustrates the flow of health data through these layers, emphasizing the collaborative nature of spacecraft health management.

The system management unit is at the helm of this process, tasked with collecting, processing, and disseminating health data. It works closely with the health data storage unit, which archives all relevant information, including health data, fault diagnoses, and system configurations. This centralized storage is crucial for data analysis, enabling the system management unit to perform statistical analyses, monitor anomalies, and maintain comprehensive records of the spacecraft's operational status.

Health data generation and processing involve synthesizing health statuses, conducting statistical analyses, monitoring for sudden changes, and reporting events. This ensures that we identify and address any potential issues promptly. Sharing health data across the spacecraft system improves accuracy, robustness, and efficiency. It keeps all subsystems informed of the overall system state and enables cross-verification for more accurate fault detection.

Health data organization and scheduling are critical for managing the vast amounts of information generated by the spacecraft. Internally, the team periodically schedules data for routine updates and dynamically schedules it for significant changes, optimizing communication bandwidth and ensuring timely delivery of information. Externally, communication with ground control employs an Advanced Orbiting

Figure 10.2 NASA's New Horizons spacecraft prepared for its launch to Pluto and its moon Charon in January 2006. It used body-mounted instrumentation for data transmission gathering and transmission of data. At Pluto's distance, radio signals took four hours and 25 minutes to traverse 4.7 billion km of space back to Earth.

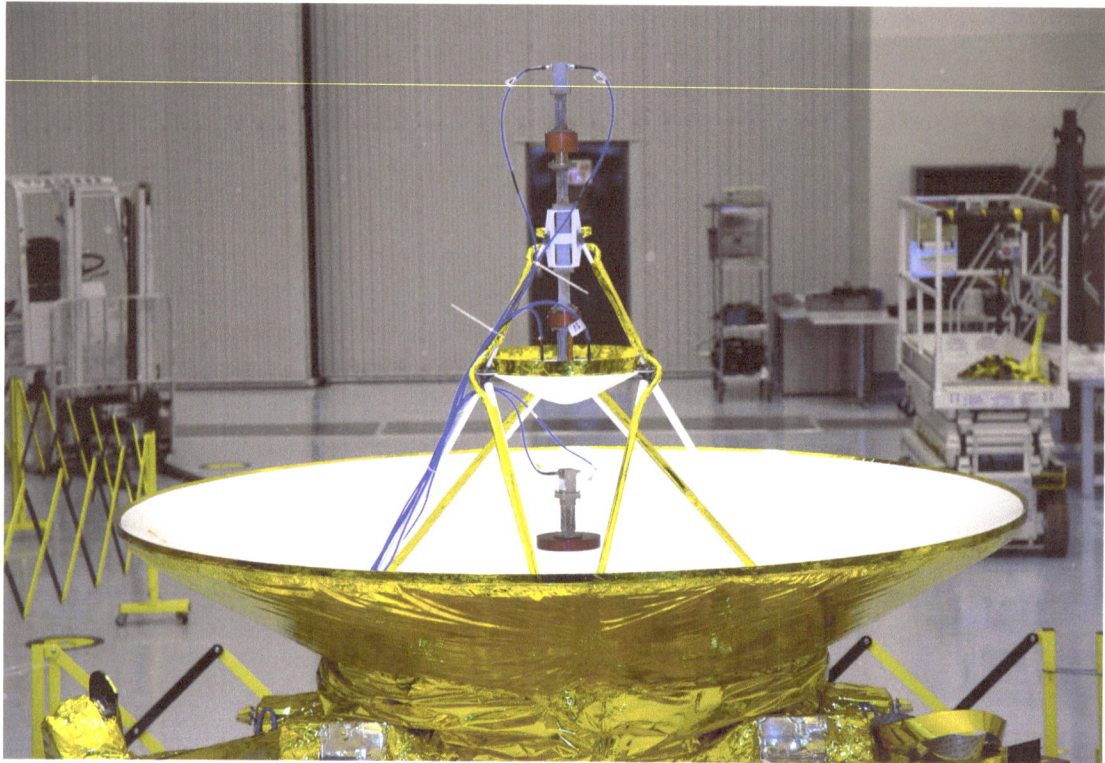

NASA

System (AOS)-based strategy, prioritizing data based on its importance and the urgency of transmission.

This structured approach to health data management—encompassing generation, sharing, and scheduling—ensures that spacecraft can maintain continuous and reliable operation. Mission controllers can enhance space mission success and safety by using detailed health data from all spacecraft levels to make informed decisions and address potential issues proactively.

10.4.
The Role of HMI and Automation

This plays a crucial role in ensuring the effective operation and safety of space missions. HMI refers to the systems that enable astronaut interaction with spacecraft controls and feedback (Table 10.3). Automation involves using automated systems to manage spacecraft functions, reducing crew workload.

Table 10.3 Reported challenges in HMI for the space domain.

Category	Challenges	Connections	Unique mission aspects
Control interfaces	• Touchscreen panels. • Manual control levers. • Voice command systems.	• Touchscreen panels offer intuitive control but may require backup manual controls. • Voice command systems can enhance efficiency but rely on integration with other control systems.	• Ergonomic design is crucial because of the confined space and microgravity environment in spacecraft.
Display systems	• Digital displays. • Head-up displays (HUDs). • Warning systems.	• Digital displays provide vital mission data, integrated with warning systems for immediate alerts. • HUDs offer real-time data without distracting from primary tasks.	• Displays must be clear and easily readable in varying light conditions experienced in space.
Automation in operations	• Autopilot for navigation. • Automated docking systems. • Robotic arm controls.	• Autopilot systems are connected to navigation controls and require precise programming. • It is necessary to seamlessly integrate automated docking systems with manual override options.	• Automation must be highly reliable because of the high-risk nature of space missions and limited intervention possibilities.
Environmental controls	• Life support. • System automation. • Temperature and humidity control. • Air quality monitoring.	• Automated life support. • Environmental controls like temperature and air quality systems closely link with systems.	• Systems must be adaptable to the unique closed environment of spacecraft and capable of handling emergencies autonomously.
Communication systems	• Internal communication networks. • External communication with ground control. • Emergency communication protocols.	• Integrating internal and external communication systems is necessary for coherent operation. • Emergency communication protocols are essential and easily accessible.	• Communication systems must be designed to function effectively despite the vast distances and potential signal delays in space.

A prime example of HMI in space is the control panel used in the ISS. These panels provide astronauts with critical information about the station's systems, such as life support, power status, and environmental controls, allowing them to make informed decisions quickly. The touchscreen controls in SpaceX's Crew Dragon spacecraft represent a more modern approach to HMI, offering an intuitive interface for crew interaction.

Due to the communication delay with Earth, the Mars rovers, such as Curiosity and Perseverance, operate autonomously, exemplifying automation. The advanced SW and sensors on these rovers allow them to navigate the Martian terrain, conduct experiments, and manage their power and thermal systems with minimal human intervention.

Inadequate HMI can lead to errors in spacecraft operation, as seen in the early days of space exploration where complex control systems sometimes led to operator confusion. In terms of automation, a failure could mean a loss of control over critical systems. An automated docking system malfunction could jeopardize the ISS and spacecraft during docking attempts.

Although no catastrophic incidents solely caused by HMI or automation failures have occurred, there have been near misses. For example, during the manual docking of Soyuz T-15 to Mir in 1986, the crew had to overcome significant challenges because of the limitations of the HMI. The HST faced a flawed mirror issue, emphasizing the significance of precise control and feedback systems in space missions.

To effectively manage aboard spacecraft, it is necessary to continuously monitor several key parameters:

- Control panel interface: Monitor user interaction logs to understand usage patterns. Check the error rate in the input to identify potential issues with the interface. Assess feedback efficiency to ensure users receive timely and accurate information.

- Touchscreen interface: Monitor touch responsiveness, screen clarity, brightness for visibility in different lighting conditions, and interface customization options. This ensures correct reaction to user inputs and user adaptability.

- Automation systems: Check system response time to ensure timely operations and autonomous navigation accuracy, especially for unmanned spacecraft like rovers, and automated system health checks for proactive maintenance.

Automation in space missions has shifted from being a supplementary feature to a core component of spacecraft operation. Autonomous systems now handle complex tasks like navigation, docking, and system health checks. The rise of AI-based capability to process vast amounts of data and provide predictive analytics has significantly enhanced decision-making processes. However, this integration presents challenges, particularly in ensuring that these systems can accurately interpret the complexities of space environments. As a result, the design philosophy for HMI systems still allows for human intervention, when necessary, as shown by the manual override capabilities in the ISS's automated docking system.

10.5.
The Role of Risk Management

Risk analysis is a critical area that involves identifying, analyzing, and mitigating risks that might affect the success and safety of space missions [10.4]. This process ensures that the spacecraft and its crew (if applicable) are protected from a wide range of potential hazards, both predictable and unforeseen, playing an integral role in risk analysis.

The Mars rover missions provide a real-world example of risk management in space. For instance, NASA's Perseverance rover team conducted extensive risk assessments before the rover's landing on Mars. Simulations and analyses were conducted to understand and mitigate risks during the rover's entry, descent, and landing (EDL) phase, also known as the "seven minutes of terror." The team prepared for various scenarios, such as parachute deployment failures and navigation system errors, to increase the mission's chances of success.

Another example is the risk management conducted by the ISS [10.5]. The ISS faces various risks, including potential collisions with space debris and micrometeoroids, system failures, and health risks to astronauts because of prolonged exposure to microgravity and radiation. The ISS program continuously monitors these risks, employing strategies like shielding for debris, rigorous maintenance schedules, and comprehensive health monitoring of the crew.

Inadequate implementation of risk management processes can lead to catastrophic consequences (Table 10.4). For instance, the Space Shuttle Columbia disaster in 2003 was a tragic outcome of underestimated risks. The shuttle disintegrated upon reentering Earth's atmosphere because of damage to its TPS, a risk that was not fully recognized before the mission. Since then, significant improvements in safety protocols and risk assessment procedures have been made within NASA and the broader space exploration community.

Table 10.4 Reported challenges in risk management for the space domain.

Category	Challenges	Connections	Not connected aspects
Technical risks	• System failures. • Design flaws. • SW errors.	• System failures can be related to design flaws. • SW errors can result from both system failures and design flaws.	• Some design flaws may not directly affect SW but mechanical or structural aspects.
Environmental risks	• Space debris. • Radiation exposure. • Extreme temperatures.	• Radiation exposure and extreme temperatures are interconnected environmental challenges. • Space debris poses a distinct risk, though it can exacerbate the impact of other environmental factors.	• Space debris collision risk is somewhat independent of other environmental factors.
Human factors	• Crew health. • Psychological stress. • Human error.	• Psychological stress directly affects the health of the crew. • Human error can result from psychological stress or health issues.	• Some aspects of human error are independent of health and stress, such as procedural mistakes.
Operational risks	• Mission planning errors. • Communication failures. • Resource limitations.	• Mission planning errors can lead to communication failures. • Resource limitations can exacerbate both planning errors and communication challenges.	• Some communication failures might be technologically induced, separate from planning or resource issues.

10.5.1.
Technical Risks

These risks can arise from system failures, design flaws, and SW errors, and their management is crucial for the success and safety of space missions. The Mars rover missions provide a notable example of managing technical risks. For instance, the Opportunity rover faced significant technical challenges because of the harsh Martian environment. Dust storms on Mars posed a risk to its solar panels, potentially leading to power failure. NASA's risk management strategy included designing the rover's panels to withstand dust accumulation and programming the rover to enter a low-power mode during storms.

Another example is the HST, which experienced a technical issue with its primary mirror. The flaw in the mirror's design, a form of spherical aberration, was a significant risk that affected the telescope's ability to capture clear images. The team successfully addressed the issue by installing corrective optics during a Space Shuttle SM, showcasing effective risk management in response to a design flaw. Improperly managing technical risks can lead to mission failure or, in the case of crewed missions, endanger human lives. The Space Shuttle Challenger disaster in 1986 is a tragic example.

The failure of an O-ring seal in its right solid rocket booster led to the shuttle's disintegration. Experts attributed this incident to a technical flaw that the risk assessment process underestimated. These examples highlight the importance of comprehensive technical risk management in spacecraft health management. Effective identification, analysis, and mitigation of technical risks are essential to ensure the functionality of spacecraft systems, the success of the mission, and the safety of crew members. Continuous monitoring, rigorous testing, and proactive planning are key strategies in managing these risks.

10.5.2.
Environmental Risks

These can significantly affect the integrity and functionality of spacecraft. These risks include exposure to space debris, harmful radiation, and extreme temperature variations, all of which must be ensured for mission success and safety. A prime example of managing environmental risks is the ISS. The ISS orbits in an environment with a high density of space debris. To mitigate collision risks, the ISS employs a debris tracking system and occasionally performs debris avoidance maneuvers. The ISS has shielding to protect against smaller debris and micrometeoroids. Radiation exposure is another critical environmental risk, especially for deep-space missions.

NASA designs Mars rovers, such as Curiosity and Perseverance, with materials and electronic components that can withstand the higher levels of radiation on Mars compared to Earth. For human missions, such as those planned for Mars, scientists consider radiation protection a significant concern and explore strategies like spacecraft shielding and limited exposure duration. Extreme temperatures in space present a substantial risk to spacecraft functionality.

The Lunar Reconnaissance Orbiter (LRO), for example, operates in the harsh thermal environment of the Moon's orbit, where temperatures can vary dramatically. The LRO uses a combination of thermal insulation and radiators to maintain optimal operating temperatures for its instruments. Inadequate management of these environmental risks can result in mission failure or cause severe damage to the spacecraft. For instance, the ESA's Envisat satellite, an EO mission, unexpectedly ceased communication in 2012. The cause of the ESA's Envisat satellite failure in 2012 remains unknown, but experts believe that space debris impact or radiation could be potential factors. These real-world examples underscore the importance of environmental risk management in spacecraft health management. Effective strategies to mitigate these risks are crucial, especially as missions venture further into deep space, where environmental hazards become more pronounced and challenging to manage.

10.5.3.
Human Factors

Human factors encompass the physiological and psychological aspects that affect astronauts during space missions. These factors are crucial to consider, as they directly affect the health, performance, and overall well-being of crew members in the unique and often stressful environment of space.

The ISS missions offer a notable example of addressing human factors. The ISS has comprehensive systems to monitor and support the physical health of astronauts, including exercise equipment to combat muscle atrophy and bone density loss because of microgravity. In addition, the ISS provides psychological support by offering communication systems that enable astronauts to stay connected with their families and mental health professionals. The Mars500 project notably studied the psychological aspect, conducting a psychosocial isolation experiment simulating a mission to Mars. The participants experienced isolation in a mock spacecraft for 520 days, which provided valuable insights into the psychological challenges and stressors of long-duration space missions. The findings have been crucial in developing strategies to support astronaut mental health for future missions.

If human factors are not adequately managed, it can lead to significant issues. On the Skylab 4 mission, the crew faced high stress and fatigue due to an intense workload and isolation. This led to a temporary refusal to communicate with mission control. This incident highlighted the importance of considering workload and psychological support in mission planning. During the early space shuttle missions, issues like space motion sickness and adaptation to microgravity were more pronounced than expected, affecting crew members' ability to perform tasks. This led to a better understanding of and subsequent mitigation strategies for these physiological responses.

These examples highlight the critical role of managing human factors in spacecraft health management. The well-being of astronauts is vital for mission success. This involves a comprehensive approach, including physical health monitoring, psychological support, and careful mission planning to manage workload and stress.

10.5.4.
Operational Risks

These encompass the challenges and hazards associated with the planning and execution of space missions. These risks can range from mission planning errors and communication failures to limitations in resources and support systems, all of which require careful management to ensure mission success and safety. The Mars rover missions provide a real-world example of managing operational risks. For instance, the precise landing of NASA's Perseverance rover on Mars required meticulous planning and execution. The mission team had to account for factors like the Martian atmosphere, terrain obstacles, and communication delays with Earth. Any miscalculation in these areas could have jeopardized the rover's safe landing and operational capabilities.

Communication is another critical operational aspect. The Apollo 13 mission served as a famous example that highlighted operational risks. After an oxygen tank explosion, the mission's aim shifted from lunar landing to safely returning the crew to Earth. The incident required rapid problem-solving and clear communication under extreme pressure, showcasing the importance of robust communication systems and contingency planning. Resource limitations also pose significant operational risks. The HST SMs involved intricate planning to ensure that astronauts could effectively repair and upgrade the telescope within the constraints of time, space-walking capabilities, and shuttle resources. Failure to manage these operational risks can lead to mission failure or, with crewed missions, endanger human lives. The Mars Climate Orbiter incident in 1999 is an example where a navigation error, because of a mismatch in measurement units between different SW systems, resulted in the spacecraft's loss. These examples underscore the importance of comprehensive operational risk management in spacecraft health management. Effective planning, robust communication systems, and contingency strategies are essential for navigating the complexities of space missions and to respond effectively to unforeseen challenges. This not only requires technical and logistical preparation, but also equipping the mission team to handle high-pressure situations and rapidly evolving scenarios.

10.6.
The Role of Life Cycle Management

Life cycle management refers to the overall process of managing a product or system throughout its entire life, including its maintenance, servicing, and support aspects. It covers activities such as regular maintenance, repairs, upgrades, troubleshooting, and customer support to ensure the product or system functions effectively throughout its lifespan. Managing the spacecraft throughout its life is a challenging task. This is because some aspects, like certain design elements or data preservation, operate more independently, with specific goals and methodologies. Table 10.5 summarizes some of these concepts.

Table 10.5 Reported challenges in life cycle management for the space domain.

Category	Concepts	Connections	Not connected aspects
Design and development	• Structural design system. • Integration SW development.	• System integration for functionality closely links with structural design. • SW development is integral to system integration, ensuring seamless operation.	• Some aspects of structural design may not directly interact with SW development.
Testing, production, and validation	• Environmental testing. • System testing. • SW validation.	• Connecting environmental testing to system testing ensures durability. • Teams often perform SW validation alongside system testing for operational integrity.	• Environmental testing can sometimes be independent of SW aspects, focusing more on physical resilience.
Launch and deployment	• Launch vehicle compatibility. • Deployment mechanics. • Initial calibration.	• Launch vehicle compatibility is crucial for successful deployment mechanics. • Initial calibration depends often on successful deployment.	• Launch vehicle compatibility has limited connection to the specifics of OBSW calibration.
EOL strategies	• Decommissioning plans. • Deorbiting procedures. • Data preservation.	• Decommissioning plans often include deorbiting procedures as a key component. • The process of data preservation ensures the saving of mission data postmission.	• Data preservation is independent of the physical decommissioning and deorbiting processes.
In-space servicing and maintenance	• Autonomous rendezvous and docking. • ORUs. • Debris mitigation. • Refueling. • Robotic manipulation.	• Capabilities of robotic arms directly influence the repairs that can be performed. • Cooperation among space agencies and private entities for standardization. • The relationship between the cost of SMs and the economic benefits of extending satellite lifespans.	• Economic viability of smaller satellites is unclear. • Most current servicing concepts focus on Earth orbit, and there is still a need to extend these services to deep space missions future goal. • Policies and regulations are still catching up with the technological advancements in space servicing.

10.6.1.
Design and Development

This stage involves careful planning and integration of various systems to ensure the spacecraft can withstand the harsh conditions of space and fulfill its mission objectives. If the design and development phase is not adequately managed, the spacecraft may fail to meet its mission objectives or experience significant problems after launch. The HST initially suffered from a flawed primary mirror, a critical design error that significantly affected its capabilities. A complex SM later corrected this issue, showcasing the importance of thorough design and development.

The JWST is a prime example. Involving intricate design choices, such as its unique sun shield and mirror structure, the development of the JWST enables it to observe distant galaxies. Integrating its systems required rigorous testing to ensure they could operate in extreme cold and a vacuum. Equally important was the development of the telescope's SW, which allowed it to process immense volumes of astronomical data. The Mars Climate Orbiter is another example where design and development issues led to mission failure. In 1999, a navigation error caused the spacecraft to be lost due to a failure to convert units from English to metric. This incident underscores the importance of precision and attention to detail in spacecraft development.

While this book does not provide a detailed analysis of this topic, it presents a basic overview of the design and development process in spacecraft engineering in Figure 10.3. These steps not only ensure the spacecraft's functionality and durability, but also lay the groundwork for successful mission operations.

Figure 10.3 Stage of spacecraft design and development process.

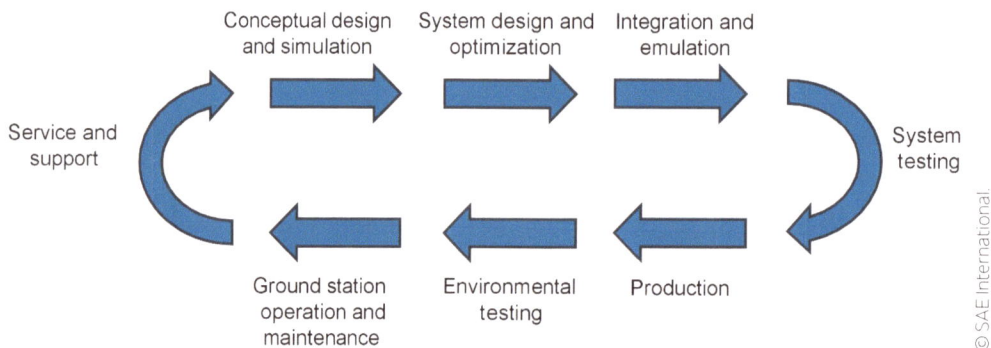

© SAE International

10.6.2.
Testing, Production, and Validation

This process involves a series of rigorous tests to validate the design, durability, and performance of the spacecraft systems under various simulated space conditions [10.6]. These would include thermal vacuum tests to simulate extreme temperatures, and vibration testing to ensure they could withstand the rigors of launch and landing. Additionally, SW validation tests subject components to ensure that their autonomous navigation systems and scientific instruments would function correctly.

An example is the JWST. Since its mission is to operate far from Earth, they subjected the JWST to intense testing. This included exposure to extreme cold and vacuum conditions to simulate the environment at its operational point in space. The telescope's components, especially its mirror and sun shield, underwent extensive testing to ensure they would deploy correctly in space.

Inadequate testing and validation can result in significant mission failures. For instance, the HST, launched in 1990, suffered from a flawed primary mirror.

Despite rigorous testing, the team did not detect the error in the mirror's shape until after the telescope was in orbit, resulting in blurred images. This caused a complex SM to install corrective optics. Similarly, the Mars Climate Orbiter was lost in 1999 due to a failure in task performance, resulting in a trajectory error and the spacecraft disintegrating in the Martian atmosphere. This incident was a stark reminder of the importance of thorough testing and validation in spacecraft development (Figure 10.4).

Figure 10.4 Typical testing, production, and validation process.

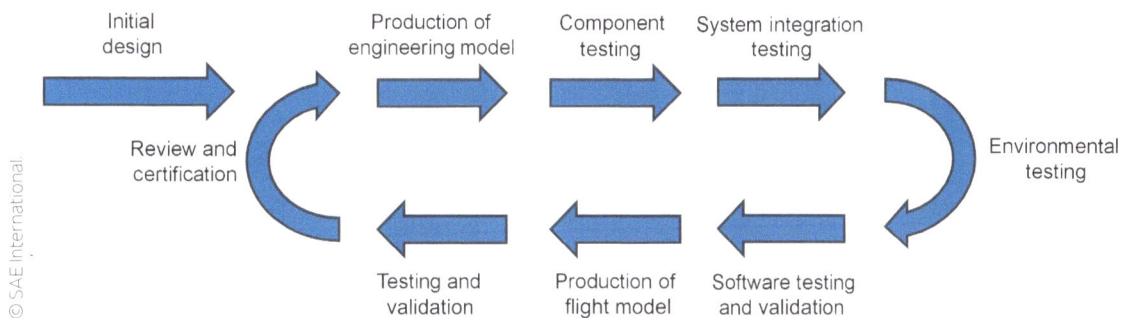

10.6.3.
Launch and Deployment

This phase encompasses the integration of the spacecraft with its launch vehicle, the actual launch, and the subsequent deployment into orbit or a specific trajectory. A notable example is the deployment of the ISS, which was launched in multiple segments and assembled in orbit. The ISS's deployment required precise coordination between various international space agencies. They carefully launched, docked, and integrated each module, forming the largest man-made structure in space. The complexity of this process was immense, considering the need for accurate alignment and connection of life support, power,

and data systems. The ISS requires constant monitoring and control from both the onboard crew and ground control centers. This includes managing life support systems, conducting scientific experiments, and performing regular maintenance and repairs. For instance, the ISS's Environmental Control and Life Support System (ECLSS), which manages the station's atmosphere, water, and temperature, requires continuous oversight to ensure the crew's safety and comfort. Regular maintenance is essential, such as fixing or replacing parts of the space station, conducting spacewalks for external repairs, and updating systems. For example, battery replacements and upgrades to solar panels have been crucial for maintaining the ISS's power supply. The consequences of failure in any of these operational aspects could be significant.

Other examples include the Mars Exploration rovers (Spirit and Opportunity) and the Challenger Space Shuttle. With the former, the deployment of the rovers involved a journey of several months to Mars, followed by a critical EDL sequence. This sequence, known as the "six minutes of terror," involved a complex series of maneuvers, including parachute deployment and airbag inflation, to ensure a safe landing on the Martian surface. Failure in the launch and deployment phase can lead to catastrophic mission failures. The Challenger Space Shuttle disaster in 1986 highlighted the risks involved in spacecraft launch and deployment, as the failure of the O-ring seal during launch caused the shuttle to disintegrate. Similarly, the Ariane 5 Flight 501 failure in 1996, caused by an SW error in the rocket's inertial reference system, resulted in the rocket's loss and its payload.

10.6.4.
EOL Strategies

Responsible decommissioning of spacecraft and minimizing their impact on the space environment and potential risks to other operational spacecraft make EOL strategies crucial. These strategies involve careful planning for the deorbiting or safe disposal of spacecraft once they have completed their missions.

A notable example of EOL strategy implementation is the controlled deorbiting of the Mir space station by the Russian Space Agency in 2001. Due to concerns about its aging systems and maintenance costs, the Russian Space Agency deliberately brought Mir back to Earth in a controlled manner after 15 years in orbit. The EOL process involved key steps:

- Preparation: The station was prepared for deorbiting, which involved removing valuable equipment and ensuring that all systems were in a safe state for the deorbiting maneuver. This included disabling certain systems to prevent any unintended interference during reentry.

- Over several weeks, the station's orbit gradually lowered with the assistance of the station's propulsion system and later a Progress cargo spacecraft attached. By using the station's propulsion system and later with the support of a Progress cargo spacecraft attached to Mir, they accomplished this.

- Controlled reentry: The last phase of the EOL process was the controlled reentry into Earth's atmosphere. This was done to ensure that remnants of Mir would land in a designated area in the South Pacific Ocean, far from populated areas. This action was taken to minimize the risk of debris causing damage or casualties.

- Impact in the Pacific Ocean: On March 23, 2001, Mir reentered the Earth's atmosphere and broke up as planned, with the debris falling into the Pacific Ocean. Russian mission control and international observers closely monitored the operation to ensure its success.

The successful decommissioning of Mir demonstrated the importance of controlled deorbiting maneuvers for large spacecraft (Figure 10.5). This contributed to the development of international guidelines and best practices for space debris mitigation and spacecraft decommissioning.

Figure 10.5 The Mir space station was the first modular space station assembled and operated in LEO from 1986 to 2001. In this photo, the Soyuz spacecraft is docked at the bottom of the Mir facility, and Progress spacecraft is on the opposite end.

NASA

Another example is the use of graveyard orbits for geostationary satellites. When communication satellites reach the end of their operational life, operators move them to a higher orbit, known as a graveyard orbit. This helps free up geostationary slots and reduces the risk of collision with active satellites. This practice has become a standard procedure in the industry. Failure to implement effective EOL strategies can lead to increased space debris and potential hazards. The 2009 collision between the inactive Russian satellite Cosmos 2251 and the operational Iridium 33 communications satellite is a case in point. This incident, which resulted in thousands of pieces of debris, highlighted the dangers of nonfunctional satellites remaining in busy orbits. The importance of EOL strategies has grown with the increasing congestion in Earth's orbit. The ESA's Clean Space initiative and NASA's Orbital Debris Program Office are examples of efforts to address space debris and promote responsible EOL practices. These strategies are vital for both sustainable space activities and safe future missions.

Such incidents have renewed interest in the mandatory disposal of defunct satellites, at the very least, moving them to a graveyard orbit. This is to reduce the risk of space debris and collisions in geostationary orbits, which are particularly crowded and strategically important for communication and meteorological satellites. As of 2024, there is no international law mandating such actions, highlighting a gap in global space governance.

10.7.
Summary

This chapter explores the remaining concepts from Chapter 5, including cybersecurity complexities and operational challenges in the space industry. It underscores the growing risk of cyberattacks resulting from increased interconnectivity and the use of older infrastructures lacking cybersecurity measures. It outlines the vulnerabilities of space stations and infrastructure to cyber threats. This emphasizes the need for strong security measures in communication, onboard systems, data protection, SW integrity, and physical system security. Operational protocols, including mission operation security procedures and emergency protocols, are crucial for maintaining mission integrity and safety. The discussion covers life cycle management, from design to EOL strategies, highlighting the importance of careful planning, testing, and validation to mitigate risks and ensure mission success. Two examples that highlight effective EOL management practices in space exploration are the controlled deorbiting of the Mir space station and the use of graveyard orbits for decommissioned satellites. These practices are hoping to reduce space debris and ensure sustainability.

References

10.1. Vollmer, B., "NATO's Mission-Critical Space Capabilities under Threat: Cybersecurity Gaps in the Military Space Asset Supply Chain," arXiv preprint arXiv:2102.09674, 2021, Retrieved from http://arxiv.org/abs/2102.09674v1.

10.2. Leveson, N.G., "Role of Software in Spacecraft Accidents," *Journal of Spacecraft and Rockets* 41, no. 4 (2004): 564-575.

10.3. Willbold, J., Schloegel, M., Vögele, M., Gerhardt, M. et al., "Space Odyssey: An Experimental Software Security Analysis of Satellites," in *IEEE Symposium on Security and Privacy*, San Francisco, 2023.

10.4. Gamble, K.B. and Lightsey, E.G., "Decision Analysis Tool for Small Satellite Risk Management," *Journal of Spacecraft and Rockets* 53, no. 3 (2016): 420-432.

10.5. Seastrom, J.W., Peercy, R.L. Jr., Johnson, G.W., Sotnikov, B.J. et al., "Risk Management in International Manned Space Program Operations," *Acta Astronautica* 54, no. 4 (2004): 273-2710.

10.6. Phillips, D.M., Mazzuchi, T.A., and Sarkani, S., "An Architecture, System Engineering, and Acquisition Approach for Space System Software Resiliency," *Information and Software Technology* 94 (2018): 150-164.

The achievements in space flight have highlighted the increasing significance of supporting maintenance and servicing spacecraft in space. Our challenges, therefore, have evolved from merely launching sophisticated spacecraft and systems to extending the life of those already in orbit. We now aim to build large structures in space to enable new scientific breakthroughs and provide systems that support space missions reliably and cost-effectively. This shift mirrors our growing awareness on Earth about reducing, reusing, and recycling resources. Abandoned spacecraft in space create hazards for newer assets and take up valuable orbital space that could be repurposed. The successful repair missions of the HST and the assembly of the ISS show the potential for maintenance support and servicing. These achievements not only assert our capability to maintain and upgrade space infrastructure but also pave the way for future space exploration. This trend is being called "New Space" as many private organizations are now entering this market with their support technologies. These acts lay the groundwork for architectures that extend the life of systems.

A study conducted at NASA's Goddard Space Flight Center concluded we could feasibly develop on-orbit satellite servicing capabilities using current technology and launch systems. This advancement would enhance our space operations, allowing us to utilize assets in near-Earth space effectively and support more distant space missions. The key activities included the evaluation of potential missions, in-space HW demonstrations, and the development of ground simulators and testbeds to help identify any technology gaps. For example,

Dextre, a robotic system, has been instrumental in improving maintenance operations on the ISS, demonstrating the potential for robotic servicing in space [11.1]. Such studies challenge common beliefs about satellite servicing, like "nothing to maintain" and "costly in-person servicing." This shows that maintenance can be economically viable, especially for satellites that play a crucial role in modern communication infrastructure. Regular maintenance can extend mission life and improve cost-effectiveness, even for satellites not initially designed with maintenance in mind. Therefore, such services will become integral to constructing large space structures, like advanced space telescopes, and supporting long-range missions with services like fuel depots. Collaborative approaches involving government, industry, and academia are necessary to develop capabilities and leverage commercial opportunities in scientific, commercial, and strategic domains. This foundation will support future ambitious projects and national interests.

11.1.1.
Why is Spacecraft Servicing Important?

Spacecraft servicing has diverse applications, each defined by its unique scientific, economic, strategic, and societal importance. For example, constructing and stocking depots for distant space travel requires support and servicing capabilities. Space robotics technologies have advanced significantly, enabling more efficient on-orbit servicing of satellites and other space assets [11.2]. Another critical service is orbital manipulation, which includes debris removal and reusing valuable orbital slots, especially in crowded orbits like LEO and GEO. The 2009 collision between an Iridium satellite and a defunct Russian satellite highlights the growing debris problem, which risks catastrophic chain reactions, known as Kessler Syndrome. Clearing failed satellites to reuse these slots is necessary because of the limited orbital space at GEO, which is crucial for communication satellites. The potential of spacecraft maintenance to extend the life of valuable assets is demonstrated by the successful life-cycle management missions to the HST. As space missions become more ambitious, the need for maintaining infrastructure becomes as essential as repair facilities for automobiles. Without such support, even well-designed systems will eventually fail. To understand the dynamics of maintaining different spacecraft, we can categorize them similarly to automotive vehicles:

- Fleet satellites: Comparable to rental cars, these are basic, high-utility commercial satellites primarily used for communication. Servicing these would focus on maintenance and refueling, with replacement being the most cost-effective option in case of major issues.

- Family satellite: Similar to privately owned family cars, these government assets offer specialized services to restrict users. Servicing might include maintenance, refueling, and some repairs to maintain strategic capabilities or bridge technology gaps.

- Exotic satellite: Like state-of-the-art cars, these are high-value, unique satellites such as the HST or reconnaissance spacecraft. Extensive repairs and maintenance for these are often justified because of their irreplaceable nature and significant capabilities.

The benefits of maintenance vary across these categories. Refueling commodity satellites must be economically viable, supporting multiple customers to extend their operational life at a cost lower than the additional revenue generated. The question of when on-orbit servicing will become an integral part of space enterprise hinges on the development of sustainable and scalable servicing technologies [11.3]. For more specialized systems, extensive repairs might be necessary to maintain specific functions. With unique, high-value satellites, any effort to extend or restore their functionality is often worthwhile.

11.2.
Debunking Misconceptions

11.2.1.
"There is Nothing to Maintain in Space"

Contrary to popular belief, extensive research shows that space maintenance is critical. A study of Earth-orbiting satellites from 1984 to 2003 reveals an average of 4.4 component failures, 3.8 systemic failures, and 0.3 deployment anomalies annually. This highlights many opportunities for servicing, particularly related to propulsion and refueling. Each year, one spacecraft needs assistance to reach its correct orbit, around 20 GEO satellites deplete their fuel, and approximately 13 undergo relocations within the GEO belt.

11.2.1.1.
Economic and Maintenance Opportunities

- Health monitoring: The active GEO satellite population (~200) could benefit from annual damage inspections and health checks. For instance, the Orion 3 Satellite's launch failure in May 1999 led to a $265 million insurance claim.

- Debris management: Removing approximately 150 nonoperational objects from GEO could reduce collision risks, benefiting all operators in the belt.

- Financial implications: Failures could result in losses exceeding $750 million annually.

- Deployment monitoring: Monitoring the deployment of approximately 20 new satellites launched into GEO each year could provide valuable services.

- Recent trends: Insurance claims for satellite failures due to propulsion leaks or incorrect orbits have reached billions of dollars recently. On-orbit maintenance and servicing could address these issues, appealing to operators and insurers. This introduces a new value proposition for satellite design and operation, offering potential cost savings and increased mission flexibility [11.4].

- Life extension: Orbital Satellite Services (OSS) estimates that over 140 commercial satellites will need decommissioning in the next decade, presenting opportunities for life extension through refueling or attaching a propulsive servicer.

- Enhanced tactical use: Refueling satellites in orbit could enable more dynamic use, such as faster repositioning within the GEO belt.

11.2.2.
"Maintenance is Expensive"

The misconception that spacecraft maintenance is prohibitively expensive stems from misunderstandings about costs. Recent research in space robotics and AI has expanded the capabilities of robotic systems, enabling more autonomous and precise servicing operations with reasonable cost [11.5]. Let us clarify this by examining three key aspects:

- Current servicing paradigm costs: The HST SMs, often cited as costing $1–2 billion each, involved complex astronaut-performed tasks. These missions, while expensive, represent extreme cases driven by high scientific value, extending the telescope's lifespan more cost-effectively than launching new observatories.

- Alternative servicing paradigms: Academic research explores cost-effective models, including servicing single or multiple satellites in similar orbits and deploying service constellations. For example, a constellation of CubeSats for GEO satellite inspection could be significantly cheaper than the HST model.

- Designing spacecraft for maintenance: Adding external HW for easier servicing and creating modular subsystems with simplified interfaces can increase a satellite's mass slightly but simplify SMs, potentially reducing overall costs. Modular design aids in both on-orbit servicing and ground processing, leading to potential cost savings.

The notion that servicing is always costly is based on limited examples. Integrating maintenance into future space architectures could lead to significant cost savings and operational flexibility. Developers can make informed decisions about the cost-effectiveness of incorporating support capabilities into their designs.

11.2.3.
"Design-for-Maintenance in Space is Compulsory"

Another misconception is that spacecraft must be specifically designed for maintenance. This belief is based on two incorrect assumptions:

- Assumption 1: Maintenance requires preplanned aids and fixtures:

  The XSS-11 mission (2005–2006) showed that reliable operations with noncooperative space objects are possible. XSS-11 conducted over 50 rendezvous maneuvers and 300 circumnavigations autonomously, proving the feasibility of intricate operations with noncooperative vehicles.

- Assumption 2: Maintaining legacy satellites is unfeasible:

  The DARPA FREND project demonstrated the capability to grapple noncooperative features like a vehicle's Marman clamp ring and bolt holes. These features, structurally robust, are suitable for docking.

  Servicing the HST proved that even parts not initially designed for servicing can be serviced with the right tools. Engineering evaluations and specialized tools enabled tasks such as refueling, HW upgrades, and repairs.

11.3.
Technology Demonstrators

The field has seen significant advancements in recent years, driven by the efforts of various major organizations in space research. These developments have been pivotal in maturing the key technologies essential for today's assembly of complex space systems. Here is an overview of some examples of technology demonstration activities presented in Table 11.1.

Table 11.1 Examples of spacecraft maintenance technology.

Canadarm3: Developed by the Canadian Space Agency, the Canadarm3 is a sophisticated robotic arm that performs a wide range of tasks in space, including the servicing and maintenance of satellites. It also assists with spacewalks. The Canadarm3 will be a critical component of the Lunar Gateway, a space station that will orbit the moon and serve as a staging ground for future missions. This advanced robotic arm is designed to operate autonomously and will play a crucial role in assembling, maintaining, and repairing spacecraft and satellites, thereby extending their operational life.

Phoenix Program: Initiated by the DARPA, the Phoenix Program focuses on reusing and repurposing components from decommissioned satellites. By salvaging functional parts and integrating them into new satellites or systems, the program aims to reduce costs and mitigate the environmental impact of space debris. This approach not only addresses the growing problem of space junk but also demonstrates the potential for sustainable practices in space operations.

RRM: A NASA initiative, the RRM aims to develop the capability to refuel satellites in orbit using robotic technology. This project successfully demonstrates the transfer of fuel to satellites, potentially extending their lifespan and reducing the need for costly replacements. By enabling in-space refueling, RRM enhances the flexibility and longevity of satellite missions, contributing to more efficient use of space assets.

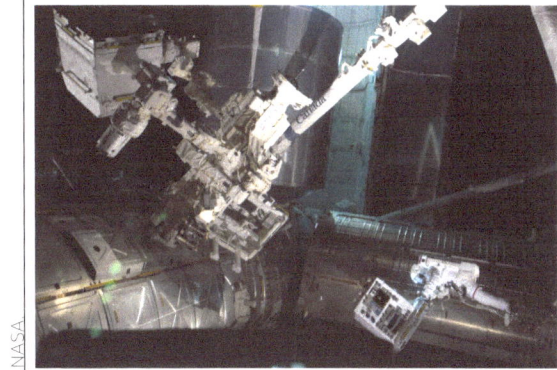

(Continued)

Table 11.1 (Continued) Examples of spacecraft maintenance technology.

Robonaut: Developed by NASA and General Motors, Robonaut is a humanoid robot designed for space exploration. It performs tasks that are too dangerous or difficult for human astronauts, such as spacewalks and repairs on the ISS. Equipped with sensors, cameras, and highly dexterous hands, Robonaut can manipulate tools and objects with precision, making it an invaluable asset for maintaining and servicing spacecraft and satellites.

NASA

SPDM, (aka Dextre): Created by the Canadian Space Agency for use on the ISS, Dextre is a robotic system with two arms, each featuring seven degrees of freedom. It is equipped with cameras and sensors for precise navigation and manipulation. Dextre performs various tasks, including maintenance, inspection, and repair of the space station. It has successfully executed missions such as replacing batteries and installing new HW, showcasing its versatility and effectiveness in space servicing operations.

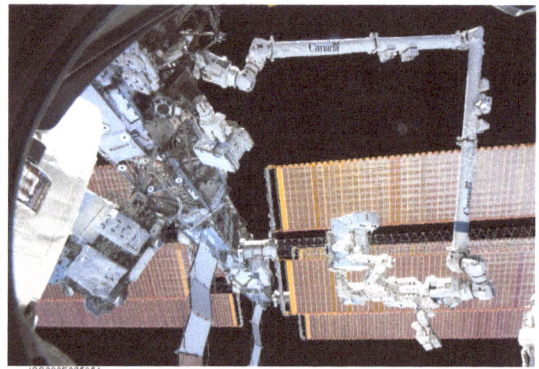

ISS026E025051

NASA

(Continued)

Table 11.1 **(Continued)** Examples of spacecraft maintenance technology.

Demonstration of Autonomous Rendezvous Technology (DART): Launched on April 15, 2005, DART was designed to demonstrate technologies for autonomous rendezvous and proximity operations. Despite encountering challenges, including a collision due to a sensor anomaly and SW issues, DART highlighted the importance of robust systems engineering in space projects. The mission provided valuable lessons for future autonomous spacecraft operations, emphasizing the need for reliable and precise control systems.

NASA

The development and deployment of these technologies highlight the growing sophistication of satellite servicing tasks. These advancements push the boundaries of what is possible in space maintenance and repair, underscoring the feasibility and complexity of such operations. The involvement of various national space agencies and private entities indicates a global interest in enhancing satellite servicing capabilities. Moreover, viable business models, such as MacDonald, Dettwiler and Associates per-kilogram fuel pricing, demonstrate the economic potential of satellite servicing. By leveraging such technologies, the Canadian aerospace company achieved greater efficiency, reduced costs, and extended operational lifespans. This not only supports current space operations but also paves the way for future explorations and sustainable practices in space.

11.4.
Assessing Technological Gaps

Contrary to popular belief, spacecraft reliability is not a given, and robotic on-orbit servicing offers a viable solution to extend their operational lifespan [11.6]. Satellite servicing technologies are now well-developed, except for autonomous operations. The focus should be on integrating existing technologies and demonstrating their performance in space. The primary challenge is effectively applying these technologies for ambitious missions, emphasizing engineering and system integration rather than new technology development. Some key areas of satellite servicing include:

- Servicer spacecraft bus: Current missions do not require new technologies for the spacecraft bus. Existing subsystems like attitude control, power, and propulsion are mature and sufficient. While improvements are possible, they are not essential.

- Rendezvous and docking: Technologies for docking with cooperative targets are mature, involving phases of rendezvous, proximity operations, and coupling. However, challenges remain with noncooperative or spinning targets, necessitating further demonstrations using technologies like cameras and LIDAR.

- Manipulation: Robotic arms are essential for tasks like refueling and component replacement. While teleoperation of large manipulators is established, autonomous control requires further validation.

- Refueling: Refueling has been demonstrated in past missions, involving fluid transfer between tanks or replacing tanks. Autonomous refueling, especially for satellites not designed for it, needs more development.

- Autonomy: Autonomy ranges from simple teleoperation to full autonomy. Higher levels of autonomy need advancements in efficiency, robustness, and capability, especially for complex tasks.

Effective integration and application of existing technologies can enhance mission capabilities while reducing costs and risks. System-level planning and integration are crucial, focusing on in-space assembly and servicing.

11.4.1.
Approaching and Servicing Spacecraft Recommendations

- Designing cooperative spacecraft: To enable autonomous rendezvous and capture (AR&C), spacecraft should be designed with features like optical retroreflectors, RF transponders, visible surface features, grapple fixtures, and proper attitude control system modes. These features have a minimal impact on mass, volume, and power requirements, but require willingness from spacecraft designers to incorporate them.

- Mode of servicing: Knowing whether servicing will be astronaut-based or robot-based is essential. Each mode requires specific design considerations like handholds and specific tooling for humans or stabilization aids and visual targets for robots.

- Design for serviceability: Historical missions, like the HST, show the importance of designing spacecraft for serviceability. This includes common connectors, standardized fasteners, accessible workspaces, clear markings, and modular design for critical functions.

- Design commonality: Industry-wide standards for spacecraft design can enhance service-ability, involving common interfaces across different spacecraft and agencies.

- Technological feasibility: The tasks required for servicing and supporting HW are feasible. Increased autonomy and robust servicing architectures are needed, using platforms like the ISS for on-orbit verification.

- Servicing legacy satellites: Servicing legacy satellites, even those not initially intended for on-orbit servicing, provides an immediate customer base and a favorable case for commercial satellites.

- Modular robotic architectures: Developing mobile, modular, reconfigurable robotic architectures is crucial for a cost-effective and upgradeable servicing infrastructure.

- Launch mass and orbit modification: The impact of launch mass and orbit modification capacity is significant. Choosing the right launch vehicles, refueling depots, and advanced propulsion systems are key considerations.

11.5.
Summary

The absence of in-space maintenance capabilities is not due to technological limitations but the need to expand the deployment of enabling capabilities. By integrating existing technologies and adopting robust systems engineering practices, satellite servicing can enhance mission capabilities, reduce costs, and mitigate risks. Collaborative efforts among government, industry, and academia are crucial to leveraging commercial opportunities and supporting ambitious space missions.

This chapter highlights the growing interest and opportunities in space flight for maintaining and extending the operational life of satellites and other space assets, reflecting an increasing emphasis on sustainability and resource conservation. Successful SMs have demonstrated the feasibility and value of in-space maintenance and repair, paving the way for future space exploration and the development of large space structures. NASA's study at the Goddard Space Flight Center supports the viability of on-orbit satellite servicing using current technologies, challenging previous beliefs about the impracticality of in-space maintenance. Regular maintenance of space assets, especially satellites crucial for modern communication infrastructure, can be economically viable, promising enhanced space operations, including extending satellite missions and supporting distant explorations. The chapter emphasizes the need for collaborative development of servicing capabilities involving government, industry, and academia to leverage commercial opportunities and support ambitious space missions. Spacecraft servicing is important for various reasons, including constructing space depots, orbital manipulation, debris removal, and reusing valuable orbital slots. With the growing problem of space debris and limited space in geostationary orbit, there is a need for effective servicing strategies. Additionally, designing spacecraft with maintenance in mind is crucial, as the right approaches and technologies can maintain even satellites not initially designed for servicing.

References

11.1. Coleshill, E., Oshinowo, L., Rembala, R., Bina, B. et al., "Dextre: Improving Maintenance Operations on the International Space Station," *Acta Astronautica* 64, no. 9-10 (2009): 869-874.

11.2. Flores-Abad, A., Ma, O., Pham, K., and Ulrich, S., "A Review of Space Robotics Technologies for On-Orbit Servicing," *Progress in Aerospace Sciences* 68 (2014): 1-26.

11.3. Hastings, D.E., Putbrese, B.L., and La Tour, P.A., "When Will On-Orbit Servicing Be Part of the Space Enterprise?" *Acta Astronautica* 127 (2016): 655-666.

11.4. Long, A.M., Richards, M.G., and Hastings, D.E., "On-Orbit Servicing: A New Value Proposition for Satellite Design and Operation," *Journal of Spacecraft and Rockets* 44, no. 4 (2007): 964-976.

11.5. Doyle, R., Kubota, T., Picard, M., Sommer, B. et al., "Recent Research and Development Activities on Space Robotics and AI," *Advanced Robotics* 35, no. 21-22 (2021): 1244-1264.

11.6. Ellery, A., Kreisel, J., and Sommer, B., "The Case for Robotic On-Orbit Servicing of Spacecraft: Spacecraft Reliability Is a Myth," *Acta Astronautica* 63, no. 5-6 (2008): 632-648.

How can we diagnose and solve the problem of a failing physical asset that is 200,000 miles away and outside direct human intervention?

This was the question asked during the critical hours following the Apollo 13 explosion, where mission control was in a race against time to safeguard the lives of three astronauts. Any misstep would have further jeopardized the compromised spacecraft. Behind the scenes, NASA utilized 15 advanced simulators, designed for comprehensive mission training, including crisis management. The Apollo 13 mission was notable because mission controllers adapted simulations to mirror the real conditions of the damaged spacecraft. This allowed them to carefully develop, test, and refine strategies for

the crew's safe return. This innovative application of simulation technology, mirroring real-time conditions, marked a significant milestone in the evolution of DTs.

The concept held significant promise for enhancing health management, aiming to boost mission success rates and utilization. Traditional methods relied on robust design and scheduled maintenance to mitigate failure risks. However, these may not fully account for the operational and environmental complexities faced in space, potentially leading to over-engineered structures or unforeseen failures. In contrast, the twin approach continuously recorded operational loads and analyzed structural damage in real time. This streamlined many processes and improved model accuracy, especially for life predictions and reusability assessments.

Today, DTs offer a transformative approach to ensuring the safety, efficiency, and longevity of spacecraft. This technology spans an array of applications critical to space missions, including condition monitoring, predictive analysis, fault diagnosis, estimation of RUL, and the development of maintenance strategies. By leveraging high-fidelity modeling and integrating data from both virtual simulations and real-world spacecraft operations, DTs enable a comprehensive understanding of spacecraft health and performance.

12.1.
What is a DT?

The structured outline in **Figure 12.1** serves as a guide to visually represent the relationships between DT-related concepts. The **digital thread** serves as the central idea, branching out into DTs and digital shadows. This highlights the seamless data flow and specific functionalities enabled by each concept. **DTs** are dynamic and can interact with their physical counterparts, offering a broad range of functionalities from simulation to control, all made possible by the digital thread. **Digital shadows** represent a more static form of digital representation, primarily focused on monitoring and analysis, with data flowing only from the physical entity to the digital model. **Digital footprints** are the data trails left by the interactions between digital and physical entities. Unlike digital shadows, which are structured representations, digital footprints can be unstructured and encompass any data generated by the use or operation of the physical counterpart. Analyzing digital footprints can provide insights into usage patterns, operational efficiencies, and potential areas for improvement.

Figure 12.1 Terminology diagram of DT-related concepts.

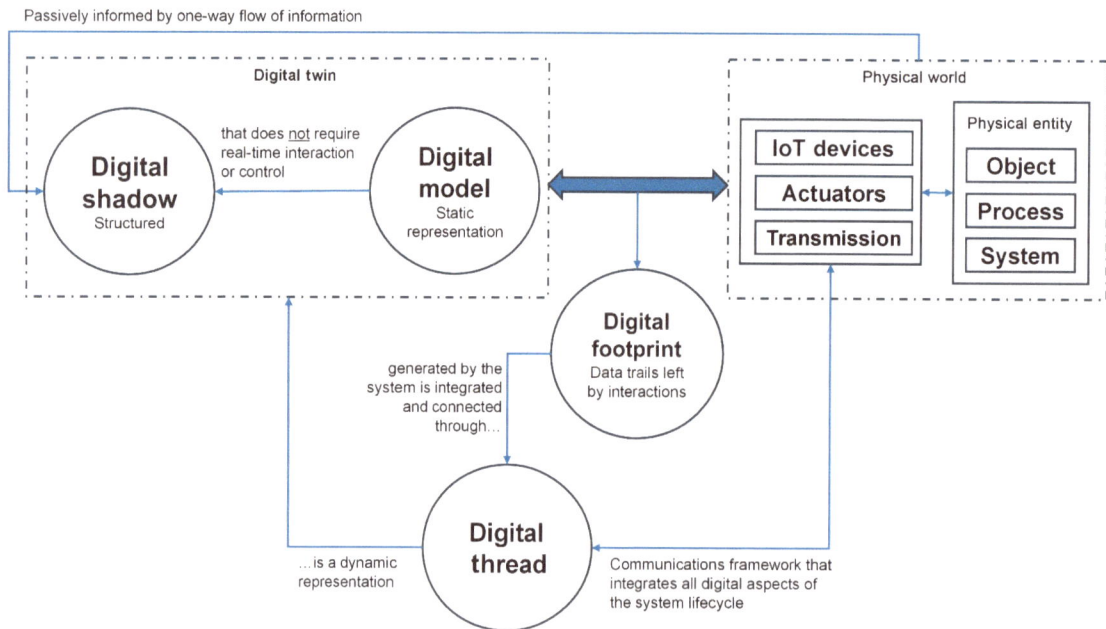

The physical and SW components are deeply intertwined, capable of interacting with each other and with humans through data exchange and control. DTs can be viewed as a component or manifestation of cyber-physical systems, providing the interface for interaction and data analysis. Having a clear understanding of these concepts offers a nuanced view of the digital representation ecosystem.

12.2.
The Role of DTs in Spacecraft Health Management

Figure 12.2a displays the initial DTs for the Apollo 13 mission. These were employed to handle the severely compromised spacecraft (Figure 12.2b) well beyond its intended operational limits. This required devising methods to conserve essential resources like power, oxygen, and water while maintaining life support for the crew and operational integrity of the spacecraft systems. Modern DTs connect a remote physical asset to its digital counterpart using continuous data flow. This enables real-time updates to the digital model in response to changes in the physical world. While the Apollo 13 mission did not utilize what we now know as "the Internet of things" (IoT), NASA leveraged cutting-edge telecommunications technology (of the time) to maintain communication with the spacecraft. These data were crucial for adjusting the simulators to accurately represent the status of the damaged spacecraft, effectively using the technology available to create a primitive form of DT. Since then, another concept of health management has gained traction as a critical framework for maintaining the integrity and functionality of advanced equipment.

Figure 12.2 (a) The Apollo Command Module Simulator at Mission Control in Houston. (b) A view of the severely damaged Apollo 13 Service Module. An entire panel of the Service Module was blown away by the apparent explosion of an oxygen tank.

NASA

(a)

(b)

NASA

DT technology plays a foundational role in health management. By leveraging historical data and real-time sensor inputs, DTs accurately reflect the entire life cycle of physical assets. This method has yielded impressive results, including creating diagnostic and predictive models for complex system faults. NASA's validation of integrating physical systems with their virtual counterparts serves as evidence for this. Other examples include the use of ultrahigh-fidelity virtual models, combined with structural and temperature models, which have enabled the prediction of aircraft structural lifespans.

12.2.1.
Differences from Previous Approaches

Several key components that differentiate DTs from traditional spacecraft management methodologies seen in previous chapters include:

- Integration vs. isolation: Unlike traditional methods that might treat data sources and system components in isolation, the DT approach integrates all relevant data and systems into a cohesive model. This allows for a more comprehensive understanding and management of the spacecraft.

- Predictive vs. reactive management: Traditional spacecraft management often relies on reactive strategies, addressing issues as they arise. On the other hand, the DT approach can predict and address potential issues before they happen.

- Dynamic vs. static modeling: Previous approaches may rely on static models that do not evolve with the spacecraft's operational life. The DT, however, is a dynamic model that updates continuously with new data, ensuring that it always reflects the current state of the spacecraft.

- Operational efficiency: The DT approach seeks to maintain the spacecraft and optimize its operations. This is a broader aim than traditional approaches, which may focus more narrowly on maintenance and repair.

12.3.
DT-Driven Framework

The DT concept faces challenges in accurately replicating complex systems because of manufacturing and material uncertainties. The goal of this concept is to create a unique representation of the physical system, utilizing multiphysics simulations and ML. By doing so, it improves adaptability to changes in conditions or operations and enhances business outcomes. However, there are still questions regarding how to accurately build models that reflect complex equipment. Additionally, establishing a dynamic link between physical assets and their digital counterparts for mutual evolution is another area of uncertainty. Furthermore, merging data from both realms to generate valuable insights is also a challenge that needs to be addressed.

The ISO13374 standard outlines the principles of open system architecture (OSA) and CBM, which form the foundation of the system-level health management framework. This standard, initially developed by Boeing, sets out the functional hierarchies within health management systems, covering informational hierarchies, sensor technologies, information processing, diagnostics, predictive analytics, and decision-making. It also defines interfaces for information exchange among various intelligent technologies, facilitating information interaction, sharing, and reusability in health management engineering. This approach has been integrated into advanced military and civilian aircraft and spacecraft systems, promoting the expansion, enhancement, and interoperability of intelligent technologies for health monitoring.

The DT framework comprises four key elements: physical devices, virtual models, services provided by virtual models to their physical counterparts, and data synchronization between the two.

In the Open System Architecture for Condition-Based Maintenance (OSA-CBM) model, a physical device captures diverse information, which is processed and analyzed by the virtual model through digital simulation, physical modeling, and data fusion. The virtual model offers health services to the physical device, supported by decision support and presentation layers. AI, ML, and data mining achieve data synchronization between physical and virtual entities. The synchronization methods are tailored to the specific needs of different physical devices, ensuring a customized approach that meets each system's requirements.

12.3.1.
Understanding of the Physical Aspects

To create a DT model, a well-defined and standardized physical system is needed. This system should be capable of spanning various real-world applications, including power systems, water supply networks, and spacecraft systems. A unique characteristic of the physical system in a DT framework is its active gathering and transmission of real-time data, which is essential for achieving autonomy.

To create a DT model for health management, you start by collecting data systematically from the beginning. Initially, we collect data on various parameters from subsystems, which then aggregate to form a comprehensive dataset representing the system-level observations. These datasets from different subsystems merge to create health indicators (HI), providing a complete picture of the system's overall health status. This bottom-up approach to data collection is crucial for the development of a DT model that accurately reflects the physical system's complexities. To effectively mirror real-world phenomena, the DT must receive a wide range of data, including operational contexts, working conditions, sensor outputs, and more. To facilitate this, the physical system must incorporate standardized data communication devices to ensure uniform data packaging and adherence to communication interfaces or protocols. These devices standardize, clean, and package various types of data across different scales within the physical system before transmitting it to the DT model in the virtual domain. This approach significantly improves the manageability and operational efficiency of data within the virtual environment, enabling a more accurate and dynamic representation of the physical system for enhanced monitoring.

12.3.2.
Constructing the Virtual Aspect

The construction of virtual equipment focuses on modeling the degradation process of physical equipment. This process is crucial for understanding how each component within a system deteriorates, influencing the overall system performance and longevity. Complex system modeling can be categorized into two primary methodologies. The first one focuses on analyzing the overall performance degradation of the system, while the second one delves into the degradation of subsystems or individual components.

The first method constructs a data-driven degradation model based on historical performance data, without considering the specific degradation paths of subsystems. This approach, exemplified by the analysis of wind farms, relies heavily on sensor data to derive performance metrics indicative of the system's degradation status. Although effective, this method demands extensive data because of the many factors affecting performance. The second method, in contrast, begins with establishing performance degradation models for each subsystem and component. It then models the system's overall performance based on the relationships between these subsystems and components, aiming to predict the system's future state. This approach requires understanding the mathematical relationships between individual component degradation indicators and the system's overall degradation indicators, requiring a detailed analysis of the interconnections within the system.

Modeling these interrelationships can follow two main paths: the first employs a fixed data model with simpler modeling requirements, focusing on parameter estimation. This includes using predefined distribution functions to model the joint distribution between degradation modes or categorizing association relationships between components into specific classes. While a large dataset is required for parameter estimation in the former, it may not accurately reflect real scenarios. On the other hand, the latter provides a more realistic representation of system conditions but is limited in depicting complex association relationships.

The second path involves modeling based on system operation principles or failure mechanisms, suitable for more complex systems. This includes hierarchical modeling, which, despite

its conceptual clarity, may not fully account for inter-level coupling degradation effects in intricate systems. Researchers have proposed advanced modeling strategies, such as bond graphs and system dynamic equations, for system-level modeling. These methods, while offering favorable predictive outcomes, depend heavily on the system's operational principles and require significant domain expertise.

It should be clear that the interest in integrating health management systems within spacecraft design is an evolving domain, marked by significant advancements and the adoption of various international standards. These standards play a crucial role in shaping the methodologies and technologies employed for fault diagnosis, health monitoring, and overall system management in space missions.

12.3.3.
Applicable Standards for Spacecraft Health Management Systems

- IEEE 1232 focuses on AI and information interactions related to test environments, setting the groundwork for interoperable diagnostic systems that can work across different platforms and vendors.

- IEEE 1451 deals with intelligent transmitter interfaces, defining communication protocols and HW specifications to ensure seamless data transmission and processing.

- IEEE 1522 complements IEEE 1232 by providing an information model that enhances diagnostics and clarifies testing objectives, improving fault detection and isolation capabilities.

- IEEE 1856 sets guidelines for developing and implementing health management systems for space missions.

- ISO 13374 outlines a comprehensive architecture for CBM, facilitating the integration of various health management functions and data analysis tools.

- OSA-CBM promotes an open architecture approach, enabling the integration of diverse health management technologies and fostering innovation in spacecraft health management systems.

These standards collectively contribute to the development of robust PHM systems for spacecraft, enhancing the reliability, safety, and efficiency of space missions. By defining clear protocols, interfaces, and testing parameters, they enable the creation of advanced health management systems capable of autonomous fault diagnosis, predictive maintenance, and efficient resource allocation. The adoption of these standards not only improves the operational performance of spacecraft but also reduces maintenance costs and extends the lifespan of space assets.

12.3.4.
Strategy of Data Interaction—Digital Thread

The interplay between physical systems and their digital counterparts is a dynamic and reciprocal process that forms the backbone of modern systems. This process involves a continuous loop of data exchange and model updating, allowing for real-time health monitoring, fault diagnosis, and predictive maintenance. It uses historical and real-time data to simulate and predict the system's behavior. The following are some model updating and data synchronization techniques:

- Bayesian algorithm: It is effective in statistical models. It offers a probabilistic representation of the degradation process through various virtual models. It excels in depicting

the uncertainty of degradation processes, enhancing the virtual model's resilience and accuracy in fault tolerance. For example, anomaly detection in cyber-physical systems using Bayesian Networks has proven to be a reliable method for identifying potential system failures [12.1].

- Random filtering algorithms: They employ techniques such as Kalman filtering, unscented Kalman filtering, and particle filtering, which are suited for dynamic system equations or state-space models. These algorithms allow the virtual model to track changes in the physical system's performance by monitoring dynamic state variables and establishing a mathematical relationship between system performance and state variables.

- ML algorithms: The integration of ML with data synchronization has revolutionized the efficiency of virtual model updates. By leveraging historical, real-time, and simulated data, ML algorithms enable the creation of a high-fidelity DT. For example, optoelectronic logic gates, implemented using photonic ICs, enable faster data processing and communication between DT models and spacecraft systems [12.2]. Additionally, DL networks are utilized for learning health data distribution and identifying failure modes [12.3].

There are other considerations of the digital thread brought about by the advancements of SW and simulation technologies. We can segment this into three distinct phases.

1. Unidirectional mapping: This refers to the initial focus of the technology, where the model information is represented in different forms and platforms. It showcases the internal dynamics and external behaviors of systems

through intuitive visualizations. Applications ranged from virtual prototypes to flight simulators. For instance, the German Aerospace Center's "Iron Bird" served as a virtual prototype for aircraft integration and validation. Despite these advancements, there is a critical limitation. The reliance on estimated or offline data for simulation analysis hindered real-time interaction and bidirectional data exchange between the digital and physical entities, marking this phase.

2. Virtual–physical interaction: Over time, simulation technology evolved into a comprehensive technical system, enabling synchronized mapping that reflects the real-time state and processes of aircraft. This development enhanced the visual experience and technical decision-making through real-time data evaluation. Technologies like dynamic data-driven simulation (DDDS) and embedded simulation exemplify efforts to incorporate dynamic real-time data into simulations. However, there was a clear need for more development in comprehensive virtual–physical interaction. Despite significant progress, challenges in deep data mining and complex data management from spacecraft systems were evident.

3. The future role: Looking ahead, the future of twins in spacecraft design and operation demands higher sophistication in data models, algorithms, and both SW and HW components. By integrating advanced models, algorithms, and ML capabilities, we expect to enable continuous knowledge acquisition, manipulation, and decision-making. This will aid in identifying potential faults by detecting discrepancies between predictive models and physical entities. The spatial dispersion and complex environments of future spacecraft

missions will pose challenges for data concentration and exchange. Yet, these challenges highlight the unique advantages of the technology in navigating and managing the complexities of space missions.

Data interaction in technology has evolved from basic mapping to virtual–physical interactions, paving the way for future advancements that will revolutionize spacecraft design, testing, and operation.

12.3.5.
Uncertainty Quantification (UQ)

UQ is a critical component of DTs in PHM and health management. It involves the identification, characterization, and management of uncertainties in the model, data, and predictions. UQ ensures that the predictions made by the DT are reliable and that the associated risks are well understood [12.4]:

- Model accuracy: Because DTs rely on mathematical models that are simplifications of reality, UQ helps in assessing the accuracy of these models and understanding the impact of simplifications and assumptions on the predictions.

- Data quality: The data fed into DTs often come with uncertainties due to sensor inaccuracies, noise, and incomplete information. UQ helps quantify these data uncertainties and their effect on the system's health predictions.

- Predictive maintenance: UQ allows for the estimation of the RUL of components with a known confidence level. This helps in planning maintenance activities more effectively and avoiding unexpected failures.

By understanding these uncertainties, decision-makers can better manage risks and make informed choices about maintenance schedules, operational limits, and design modifications. There are a few ways to handle the uncertainties in DTs:

- Probabilistic methods:
  - Monte Carlo simulation: Involves running many simulations with varying inputs based on their probability distributions to assess the impact of uncertainties.
  - Bayesian inference: Updates the probability of a hypothesis as more evidence or information becomes available.
- Nonprobabilistic methods:
  - Interval analysis: Uses intervals to represent uncertain parameters and calculates the bounds of the output.
  - Fuzzy logic: Represents uncertainty with degrees of membership rather than precise values, suitable for handling vague or imprecise information.
- Hybrid methods:
  - Stochastic collocation: Combines the advantages of polynomial chaos expansions and sparse grid techniques to quantify uncertainty in complex models efficiently.
- Sensitivity analysis:
  - Determines how the variability in the output of a model can be assigned to different sources of uncertainty in its inputs.

12.4. Evolution of Spacecraft DT Technology

The development of DT technology in design and testing has evolved significantly, building on the foundation of simulation techniques to create a sophisticated 3D model. We can categorize this evolution into three main phases:

1. Physical testing phase: Initially, spacecraft design relied heavily on physical simulations, using actual objects or scale models for tests like wind tunnel and free flight tests. These methods offered direct visual feedback and high reliability. Despite advancements in simulation technology, physical testing remains a crucial, albeit costly, method because of its effectiveness in responding to and identifying potential damages or defects without real-time sensor data. At this stage, the DT concept is more of a physical replica, lacking instant data exchange but capable of replicating environmental conditions to reflect the physical entity's status.

2. Semiphysical simulation: This phase introduced a blend of physical components and simulation, creating a closed loop that simulates the performance of spacecraft components, such as guidance systems and onboard equipment. It marked a departure from purely physical simulations, employing hierarchical modeling to tackle complex physical problems. Notably, during the Apollo program, NASA used semiphysical simulators for astronaut and mission controller training, establishing a weak link between physical and virtual models. This "semi-twin" phase allowed for analog signal-based simulations, bridging the gap between physical and virtual realms.

3. Computational simulation phase: The transition to computational simulation represented a significant shift, with digital models gradually taking over physical components. This phase enabled a one-directional data flow from physical to virtual entities,

allowing for detailed analysis of external loads and structural responses within a digital environment. Although this approach facilitated communication between physical and virtual models, it also highlighted the limitations of modeling physical problems, sometimes leading to inaccuracies.

12.4.1.
A Conceptual Model

The conceptual model of a spacecraft represents a sophisticated framework designed to integrate comprehensive data across physical and virtual dimensions, enhancing spacecraft design, simulation, and operational efficiency (Figure 12.3). Five critical layers can structure this model:

- Physical asset: The foundation of the DT system, the physical entity layer, encompasses the actual spacecraft and its components. It involves detailed analysis and modeling of the spacecraft's subsystems and their interactions, facilitated by distributed sensors and embedded systems to capture multisource data. This layer ensures real-time data transmission to the virtual entity, enabling accurate replication and analysis in the environment. For example, an IoT-based fault diagnosis systems, when combined with DTs, provide real-time monitoring and predictive maintenance capabilities, enhancing spacecraft operations [12.5].

Figure 12.3 A model for spacecraft DT.

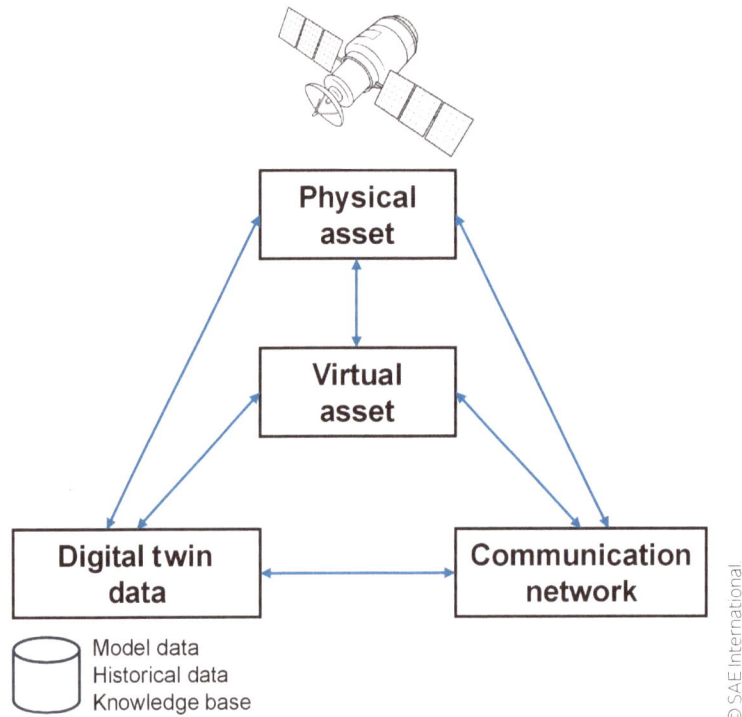

- Virtual asset: This layer serves as the digital counterpart of the physical spacecraft, comprising geometric, physical, motion, and state models. It dynamically adapts to changes and challenges through model reconfiguration and modification, addressing specific scientific and operational issues. Tools like ADAMS, STK, and MATLAB support various analyses, from structural to orbital dynamics, enriching with detailed simulations and real-time operational visualizations.

- DT data: At the core of the system, the data layer amalgamates information from the physical and virtual entities, including specifications, performance metrics, and environmental conditions. It extends to encompass satellite network data and derived analytics for simulation, prediction, and decision-making. This layer is pivotal for integrating expert knowledge, algorithms, and historical data, facilitating comprehensive information sharing and value generation [12.6].

- Communication network: Extending the physical entity into the digital realm, this layer connects the spacecraft's network nodes, enabling data exchange and information transmission across the system. It evolves from simple data exchanges to complex data fusion, linking diverse satellites into a coherent network. This layer supports the exchange of relay information, positioning data, orbit parameters, and control commands, which are crucial for cluster cooperation and mission execution.

- Integration and Interoperability: The twin model emphasizes the importance of scalability and interoperability within the communication network to meet the unique demands of space operations. There needs to be a progression from individual

spacecraft analysis to cluster system iterations, addressing the complexities of multientity interactions and data sharing. This model not only facilitates the convergence of all available data, eliminating information asymmetries, but also enables an interconnected and logical scene for enhanced spacecraft services.

12.4.1.1.
Data Sources

The digitization of mission operations shows the importance of data collection, storage, and analysis. With advances in embedded sensors, low-power wireless communication, and efficient signal processing technologies, a data-aware approach is facilitated for constructing virtual systems across various application scenarios. Since spacecraft structures are complex, many devices and systems functions like attitude change, orbit control, and environmental monitoring need to be monitored. Key data types essential include:

- Flight status data: This encompasses data on the spacecraft's position, acceleration, speed, and orbit parameters postlaunch, providing a comprehensive view of its flight status.

- Survival environment data: These data points monitor conditions such as temperature, air pressure, humidity, and gas concentrations, which are crucial for ensuring suitable conditions for both astronauts and equipment.

- Flight attitude data: Attitude parameters, determined by sensors like solar and star sensors, are vital for mission tasks, including rendezvous, docking, and orbit correction. Attitude parameters, represented through Euler angles, unit quaternions, or rotation matrices, depict these parameters.

- System temperature data: We monitor operational temperatures for onboard equipment to ensure they remain within optimal ranges. This involves adjusting temperature control systems based on collected data to maintain a heat balance within the spacecraft.

- Working state data: These include operational and performance parameters for systems, subsystems, and devices. We can categorize parameters based on their stability or variability over time, including constant, monotonically changing, periodically changing, or situationally changing parameters.

- Visual perception data: Visual data, obtained through various imaging technologies, support tasks like attitude measurement and on-orbit maintenance. These data also aid in astronaut monitoring and environmental assessment within the spacecraft.

12.4.1.2.
Twin Configuration

Spacecraft DT systems face the added challenge of navigating the complexities of space. They must evolve from single-function systems within a spacecraft to sophisticated networks that enable communication and collaboration among multiple spacecraft. This evolution marks a transition from individual entity intelligence to collective swarm intelligence. In this case, a systematic configuration can be used to tackle challenges specific to the environment, including cross-space cooperation, data interaction, and network evolution [12.7]. Key components include:

- Collaborative systems integration: Within a spacecraft, the collaboration between various systems is crucial for data integration, subsystem planning, and intelligent decision-making. This requires a configuration

structure that can integrate multidimensional data and translate scattered data into interdisciplinary models.

- Model data collection and reusability: The need to provide a reusable and extensible collection of model data. This collection evolved with the spacecraft's on-orbit service, simulating the physical entity to offer insights into system behavior, performance evaluation, and quality measurement.

- Health correlation model: A foundational aspect is establishing a health correlation model between the spacecraft's perception systems and the physical satellite. This involves describing and correlating data with satellite business scenarios based on multi-source information perception.

- Multidisciplinary: The DT can use a multidisciplinary simulation model to mathematically represent various business problems. This model predicts actual issues through modular simulation, using perception data as a source.

- Integration of physical and virtual models: The integration of physical information and virtual models with both local and cloud data platforms can be useful. Intelligence can support this integration and computing systems to provide comprehensive information and decision support.

- Service provision: The system can offer services such as system state prediction, collaborative simulation for satellite clusters, information network monitoring, and satellite behavior prediction. These services support satellite body state changes, orbit model control, and mission completion for satellite cluster networks.

12.4.1.3.
Information Network Model

The model is expected to be dynamic and have a distributed structure for processing and managing vast amounts of space mission data. It can use a hierarchical integrated network for efficient allocation of storage and computing resources. This network evolves as data accumulates from various sources, including local and cloud services, ensuring seamless integration across multiple dimensions. Some key components of the information network can include:

- Local data service:

  - Spacecraft subsystems generate vast amounts of data from complex missions, necessitating constant updates to onboard data collection systems.

  - Examples include the MRO, which sent back 25 TB of data over seven years, highlighting the exponential growth of remote sensing data.

  - Big data applications in aerospace, like those by Orbital Insight, leverage satellite data for innovative applications across multiple sectors.

  - The system should be able to integrate human knowledge with data-driven models, improving performance and providing a comprehensive data source.

- Cloud data service:

  - Limited satellite resources pose a challenge to data computing and transmission efficiency.

  - Cloud data services treat satellites as integrated computing and storage nodes, using virtualization to manage resources.

- Initiatives by companies like Cloud Constellation, IBM, and SpaceChain demonstrate the potential for cloud technology and AI in space, enhancing data processing and storage capabilities.

- Multi-spacecraft information network:

  - This network facilitates information sharing and resource allocation through satellite interconnectivity and satellite-to-ground communication.

  - Projects like OneWeb, SpaceX, and China's Hongyan and Hongyun aim to create robust satellite constellations for global communication, showcasing the network's resilience and mission robustness.

- Ground data center:

  - Acts as an extension of the space-based distributed data system, offering unified resource management for the spacecraft information network.

  - Uses big data storage architectures and hybrid database management strategies for handling complex, distributed data.

  - NASA's Nebula and Integrated Rule-Oriented Data System (iRODS) platforms exemplify the use of big data technology in aerospace, processing data streams from lunar photos and Mars missions.

The integration and evolution of this model is a comprehensive approach to managing spacecraft data. It involves integrating local and cloud data services with a multi-spacecraft information network and ground data centers. This model supports the entire life cycle of spacecraft missions, from data collection and processing to decision-making and mission execution. By standardizing and sharing satellite remote sensing data, the model aims to enhance the compatibility and interoperability of satellite services, paving the way for more efficient and collaborative space missions.

12.4.2.
Toward a Smart Spacecraft

The concept of "smart" in spacecraft, developed from extensive data analysis and computation, is critical for enhancing on-orbit operations. Achieving intelligence involves two critical capabilities: independent problem-solving using interdisciplinary knowledge and self-learning from historical mission data. DTs embody these capabilities, simulating life cycle evolution, data models, and independent learning, laying the groundwork for spacecraft intelligence.

Future development of spacecraft DT with intelligence includes:

- Autonomous cognition:

  - Autonomous cognition enables spacecraft to perceive their state and external environment in real time, completing missions autonomously without human directives. This includes navigating challenges like orbital transfers, interplanetary flight path planning, collision avoidance, and on-orbit servicing. Autonomous systems will play a pivotal role in the next generation of space exploration, addressing challenges such as system failures and environmental hazards [12.8].

  - The process involves the spacecraft's CPU analyzing mission parameters from vast data sets, incorporating knowledge from orbital mechanics, dynamics, and control algorithms into the twin. This systematic simulation predicts optimal flight paths and

executes missions while updating the spacecraft's state in the virtual model in real time, forming a closed-loop control system.

  - "Justin," a robot developed to showcase autonomous cognition, performs complex tasks under human supervision in orbit. Additionally, computer chips for CubeSats exhibit autonomous cognition by autonomously processing satellite images.

- Unmanned autonomous cognition: This aspect focuses on the spacecraft's ability to recognize and analyze problems independently, leveraging computer science, mathematical physics, materials science, and other disciplines. It encompasses the spacecraft's capacity for spatial adaptability and behavioral autonomy, crucial for efficient mission completion.

- Autonomous operation and maintenance: Spacecraft are expected to manage their operation and maintenance. They learn from similar historical missions to update and perfect their knowledge base dynamically. This self-learning capability ensures continuous improvement and adaptability of the spacecraft's operational strategies.

- Collaboration of heterogeneous clusters: The intelligence of spacecraft also extends to the collaborative operation of heterogeneous clusters, enabling coordinated missions among diverse spacecraft systems. This collaboration enhances mission efficiency and resource utilization across the spacecraft fleet.

- Autonomous operation and maintenance: These are critical for spacecraft to maintain optimal functioning over extended periods, especially since any malfunction can lead to significant maintenance costs. This autonomy aids spacecraft in swiftly recovering from system failures, preventing catastrophic outcomes. Some examples of advances:

- Pioneered the use of intelligent fault diagnosis in spacecraft, with the Gemini spacecraft featuring a fault detection system for monitoring critical parameters.

- The US and Germany have conducted significant research on spacecraft formation flight, with projects like TanDEM-X, TerraSAR-X, and NetSat demonstrating autonomous satellite formation flying and coordination.

- China: Established a laboratory dedicated to on-orbit fault diagnosis and maintenance, applying these technologies to various space missions.

- Mission architecture: Agencies like ESA and NASA have utilized satellite constellations for collaborative missions.

12.5.
Examples

12.5.1.
Structural Health Management of Reusable Spacecraft

Ye et al. present a comprehensive application of DT technology aimed at enhancing the reliability and efficiency of spacecraft maintenance and operation [12.16]. The development and implementation of a DT for structural health monitoring involve a multifaceted approach. It integrated real-time simulation techniques with online monitoring data to create a dynamic,

predictive model of the spacecraft's structural health. The core idea lies in bringing together diverse models, multisource data, and uncertain parameters into a DBN that has shown promise in managing prognosis and uncertainty. For example, Latin hypercube sampling can offer a robust method for propagating uncertainty in complex systems, providing valuable insights for spacecraft health management [12.9]. However, some challenges remained in dealing with the inherent uncertainties in fatigue parameters and the limitations of certain inference algorithms. This was addressed by using a generalized inference algorithm to improve health condition predictions for spacecraft. Figure 12.4 shows the proposed DT. It is structured into two main stages—offline and online—each comprising four distinct modules tailored to support the spacecraft throughout its life cycle as listed in Table 12.1. The twin depicts communication between modules, some with a bidirectional flow to enhance model accuracy and fault mode understanding. This ensures that both stages contribute valuable information, from fault mode identification to real-time structural health monitoring. The framework aimed to distinguish between DT functions (such as diagnosis, model updating, performance evaluation, and data storage), and applications (such as monitoring, health evaluation, prognosis, and data sharing). This introduces a distinction between the development, implementation, and maintenance phases, aligning with spacecraft operational scenarios from ground preparation to in-orbit and return phases.

Figure 12.4 DT framework for reusable spacecraft.

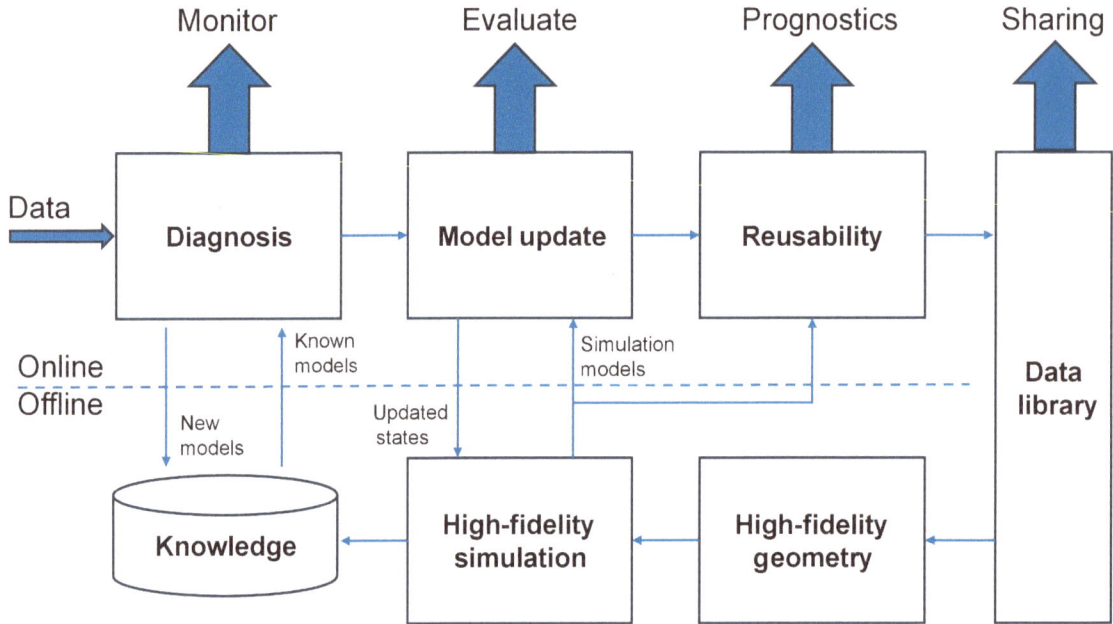

Table 12.1 Various modules tailored to support a spacecraft DT.

Offline stage modules	Online stage modules
High-precision geometric model module: Incorporates the original CAD model, updated with as-manufactured characteristics and damaged or defective features. This dynamic adaptation accounts for manufacturing discrepancies, ensuring the geometric model accurately reflects the spacecraft's physical state.	Diagnosis module: Uses real-time monitoring data and intelligent algorithms for fault detection, identification, and evaluation. This process pinpoints fault locations, types, and severity, informing the subsequent model updating process.
High-precision simulation model module: Encompasses multiphysics and multisystem simulations, employing reduced order and surrogate modeling to enhance computational efficiency without sacrificing accuracy.	Model updating module: Focuses on UQ and DBN integration for model refinement. This module uses monitoring data to update simulation models dynamically, improving their accuracy and relevance.
This module integrates various subsystems, addressing the complex environmental interactions faced by spacecraft.	Reusability evaluation module.
Knowledge library module: Contains information on fault and failure modes, along with corresponding strategies for mitigation. This repository aids in diagnosing issues like solar panel unfurling failures or structural fatigue, providing actionable solutions.	Analyzes performance degradation, residual life, and reliability to assess the spacecraft's readiness for subsequent missions. Updated state parameters improve evaluation precision, guiding maintenance and mission decisions.
Offline data library module: stores preflight data, including sensor history, operational records, and model-related information. This module facilitates data interaction across the offline stage, allowing for continuous data access and updates.	Online data library module: Collects and stores in-flight data, including sensor monitoring and real-time state parameters. Like its offline counterpart, this module supports data exchange with other online modules, ensuring comprehensive data availability.

12.5.2.
Spacecraft DT for On-Orbit Operations

In their study, Yang and Li used DT technology to simulate spacecraft operations, with a specific emphasis on four-dimensional models for data acquisition and system configuration [12.10]. The main subject of this paper is the utilization of DT technology for on-orbit spacecraft. The authors introduce the concept of spacecraft digital twin (SDT) and examine four stages of simulation development using the DT approach. Throughout its life cycle, the SDT framework incorporates virtual and physical spaces to optimize spacecraft operation and management. Some essential elements of the methodology are:

- Four-dimensional model: A conceptual structure designed to adapt spatial distribution, which encompasses data acquisition, system configuration, and data service models.

- Data acquisition: The process of gathering real-time data from various spacecraft sensors and subsystems.

- System configuration: Establishing a virtual model that mirrors the physical spacecraft, enabling continuous monitoring and simulation.

- Data service model: Implementing a service-oriented architecture to manage and distribute data efficiently.

The implementation of the SDT framework incorporates a hierarchical design scheme to achieve robustness and reliability. It involves the development of comprehensive virtual models that are continuously refreshed with up-to-date data from the spacecraft. In addition, the implementation includes the integration of advanced technologies like ML algorithms and blockchain to enhance data security and sharing

capabilities. The paper emphasizes the wide-ranging applications of DT technology, extending beyond spacecraft to fields like aerospace, industrial manufacturing, and smart cities. To fully achieve the benefits of DT in on-orbit operations, the authors highlight the crucial role of continuous innovation and integration of new technologies.

On-orbit management is set to undergo a revolutionary transformation with the integration of DT technology, allowing for autonomous cognition, operation, and maintenance. It improves mission success rates by enabling intelligent decision-making and reducing operational risks. The ability of the SDT framework to simulate real-time conditions and anticipate potential problems is essential for ensuring spacecraft health and prolonging its operational lifespan.

12.5.3.
Intelligent Monitoring System for Vacuum Thermal Tests

Wu et al. introduced a DT framework for intelligent spacecraft monitoring during vacuum thermal tests [12.11]. The system combines HW and SW to collect data comprehensively, allowing for advanced analysis, fault detection, and optimized decision-making during testing. Leveraging a DT framework, this study presents an intelligent monitoring system for conducting spacecraft vacuum thermal tests. With its more reliable and intelligent control system, the framework aims to boost testing efficiency and decrease costs. This system effectively handles thermal characteristics and temperature control in spacecraft components using features like one-key boot, thermostat technology, parameter identification, and self-tuning control.

The methodology includes:

- DT model mapping: Constructing a DT model that accurately reflects the real-world spacecraft vacuum thermal test environment through a hierarchical design scheme.

- HW and SW integration: Designing a system that integrates both HW (sensors, data acquisition devices), and SW (analysis algorithms, control systems) to reliably capture dynamic data during tests.

- Data analysis and fault detection: Utilizing the DT framework to analyze testing data, detect faults, and optimize decision-making processes.

The implementation of the system requires the development of a comprehensive DT model that accurately represents the vacuum thermal test process. It combines state-of-the-art sensor technologies for precise data collection and employs advanced SW algorithms for analyzing data and detecting faults. Integration of these components enables thorough monitoring and control of the testing environment. Testing efficiency is greatly improved, and costs are reduced with the implementation of this intelligent monitoring system. By incorporating real-time data analysis and fault detection, the system enhances the dependability and precision of thermal testing for spacecraft components. As a result, decisions become more informed, testing errors decrease, and overall test reliability improves.

The paper showcases a real-life experiment to prove the effectiveness of the intelligent monitoring system, showing how the system can be applied in real-world scenarios to achieve effective testing outcomes.

12.5.4.
Dynamic Reliability Prognosis for Reusable Spacecraft

In their paper, Gao et al. propose a dynamic reliability prognosis method using a DT framework [12.12]. This method allows for adjusting maintenance intervals, predicting structural failures, and planning missions with quantified risks for reusable spacecraft. Utilizing advanced simulation techniques, like Monte-Carlo simulations, the framework can predict reliability by taking into account individual flight histories and nonconstant unit hazard rates. By using this method, the decision-making process in mission planning is improved and the structural integrity of reusable spacecraft is ensured, leading to higher mission success rates and decreased operational risks. Specifically tailored for reusable spacecraft, the paper introduces a dynamic reliability prognosis method based on a DT framework. Key components of the methodology consist of:

- DBN: Used for integrating uncertainties and dynamically updating the model.

- Maintenance point setting: Maintenance is triggered when predicted structural reliability falls below a threshold or unexpected conditions (e.g., landing impact) occur.

- Data assimilation: Inspected data are continuously assimilated to update the structural reliability model.

The implementation of the framework includes the creation of a DT model with a DBN for predicting reliability. The method's effectiveness is demonstrated through a simplified spacecraft structure example, which includes fatigue load and landing impact. Continuous model updating guarantees precise and dependable predictions.

The real-time adjustment of maintenance intervals, early warning of structural failures, and quantified risk allow for dynamic reliability prognosis with this method. It significantly increases the reliability and safety of reusable spacecraft, potentially prolonging their lifespan and reducing maintenance expenses.

The paper underscores the importance of choosing the correct landing impact probability for accurate maintenance interval control in reliability prediction. The proposed method's capacity to dynamically update reliability predictions based on new data is a ground-breaking development in spacecraft mission planning and maintenance strategies.

12.5.5.
Orion DT Pilot Project

As a part of their Digital Transformation initiative, NASA initiated the Orion DT pilot project in 2020. The project aimed to fill the knowledge gap in Orion subsystem design for Artemis I, strengthen model-based system engineering (MBSE) capabilities, and provide hands-on training for NASA personnel [12.13-12.15]. The methodology involves:

- Systems Modeling Language (SysML) modeling: Developing an executable SysML model that integrates with the physical assets of the Orion spacecraft.
- DT framework: Creating a DT of the Orion spacecraft to simulate its subsystems and provide real-time data for decision-making.
- Data integration: Utilizing available system information from the mature Orion program to build a comprehensive digital model.

The project used a structured approach to develop the SysML model and DT, with a focus on:

- System information gathering: Collecting detailed information from the Orion program to inform the digital model.
- Model development: Building the SysML model to accurately represent the spacecraft's subsystems and their interactions.
- Integration with physical asset: Ensuring the DT is linked to the physical Orion space-craft, allowing for real-time updates and simulations.

The Orion DT has greatly cut down engineering question response time and reduced human resource requirements compared to traditional methods. The ability to do this is extremely important for upcoming missions, especially when it comes to managing spacecraft complexity and making quick decisions in critical or emergency situations. The paper summarizes a number of important lessons learned from the project:

- Importance of early integration: Integrating DT technology early in the spacecraft design process is crucial for maximizing its benefits.
- Training and knowledge transfer: Providing hands-on training for NASA employees in MBSE and DT methodologies is essential for successful implementation.
- Scalability and reproducibility: Developing a methodology that can be scaled and applied to other NASA programs and missions.

The Orion DT project showcases how DT technology can improve spacecraft design, operation, and maintenance. The DT enhances decision-making, minimizes risks, and boosts mission success rates by simulating the space-craft in detail and in real time. The paper highlights the importance of ongoing innovation and incorporating new technologies to maximize the advantages of DTs in space exploration.

12.6.
Summary

The concept of "twin" in NASA's Apollo program, particularly during the Apollo 13 mission, can be seen as a precursor to modern DTs. NASA used high-fidelity simulators, which, while not digital in today's sense, functioned similarly by replicating the spacecraft's conditions for problem-solving. These simulators allowed mission control to test and refine strategies in real time, effectively acting as physical twins of the spacecraft. This innovative use of simulation technology was crucial for the safe return of the Apollo 13 crew and marked an early instance of the twin concept in complex system management.

DT technology provides a transformative solution for spacecraft safety, efficiency, and longevity. It supports condition monitoring, predictive analysis, fault diagnosis, and maintenance strategy development through high-fidelity modeling and real-time data integration. This approach not only enhances mission planning and execution but also provides a platform for immersive training, improving crew preparedness for complex scenarios and maintenance tasks in space. This chapter introduces a DT framework aimed at improving the health management of spacecraft, leveraging real-time data and predictive analytics to enhance mission success and efficiency. Traditional approaches rely on scheduled maintenance, which may not fully capture the complexities of space operations, potentially leading to over-maintenance or unexpected failures. With a DT approach, we can address these issues during the design stage. This involves the continuous monitoring of operational loads, analysis of structural damage, optimization of maintenance processes, and improvement of life prediction models for spacecraft. This technology enables precise risk assessments and mission adjustments, reducing unnecessary maintenance and cutting costs.

References

12.1. Nanduri, A., Wang, L., and Shankar, R., "Anomaly Detection in Cyber-Physical Systems Using Bayesian Networks," *Engineering Fracture Mechanics* 231 (2020): 107076, doi:https://doi.org/10.1016/j.engfracmech.2020.107076.

12.2. Lim, S.T., Ge, S.S., and Lee, T.H., "Optoelectronic Logic Gates Using Photonic Integrated Circuits," *Proceedings of SPIE* 12100 (2022): 1210008. https://dx.doi.org/10.1117/12.2645271.

12.3. Zhang, J., Liu, Z., and Li, Y., "Condition-Based Maintenance of Industrial Assets Using Deep Learning," in *IEEE International Conference on Prognostics and Health Management (PHM-Yantai)*, Yantai, China, 2022, 9942083, https://dx.doi.org/10.1109/phm-yantai55412.2022.9942083.

12.4. Smith, R.C. et al., *Uncertainty Quantification: Theory, Implementation, and Applications* (Philadelphia: Society for Industrial and Applied Mathematics, 2014).

12.5. Zhang, Y., Wang, H., and Liu, X., "Development of a Novel IoT-Based Fault Diagnosis System for Smart Manufacturing," in *IEEE International Conference on Electrical and Control Engineering (ICECE)*, Xi'an, China, 2022, 10048639, https://dx.doi.org/10.1109/ICECE56287.2022.10048639.

12.6. de Souza, R. and Lima, P., "Enhancing Manufacturing Process Efficiency through Digital Twin and Machine Learning Integration," *Computers in Industry* 148 (2023): 103948. https://dx.doi.org/10.1016/j.compind.2023.103948.

12.7.   Li, X., Jiang, S., and Chen, J., "A Review of Intelligent Fault Diagnosis Methods for Rotating Machinery Based on Deep Learning," *IEEE Access* 9 (2021): 121376-121396. https://dx.doi.org/10.1109/ACCESS.2021.3100683.

12.8.   Anderson, M. and Roberts, T., "Autonomous Systems for Space Exploration: Innovations and Challenges," *IEEE Aerospace Conference* 2023 (2023): 10115665. https://dx.doi.org/10.1109/AERO55745.2023.10115665.

12.9.   Helton, J.C. and Davis, F.J., "Latin Hypercube Sampling and the Propagation of Uncertainty in Analyses of Complex Systems," *Risk Analysis* 19, no. 1 (2003): 102-120.

12.10.   Yang, W. and Li, S., "Application Status and Prospect of Digital Twin for On-Orbit Spacecraft," *IEEE Access* 9 (2021): 106489-106500, doi:https://doi.org/10.1109/ACCESS.2021.3100683.

12.11.   Wu, D., Li, Z., Sun, J., and Zhu, L., "Digital Twin Based Intelligent Monitoring System for Spacecraft Vacuum Thermal Tests," *Proceeding of SPIE* 12257 (2022): 122570T-1-122570T-7.

12.12.   Gao, B., Ye, Y., Pan, X., Yang, Q. et al., "A Dynamic Reliability Prognosis Method for Reusable Spacecraft Mission Planning Based on Digital Twin Framework," *Journal of Nondestructive Evaluation* 42, no. 3 (2023), doi:https://doi.org/10.1007/s10921-023-00889-1.

12.13.   National Aeronautics and Space Administration (NASA), "Orion SysML Model, Digital Twin, and Lessons Learned for Artemis I," 2023, Retrieved from NASA Official Website.

12.14.   Coble, J.B. et al., "Prognostics and Health Management for Space Systems Using Digital Twin Technology," NASA Technical Reports, 2018, Retrieved from NASA Technical Reports Server.

12.15.   Pierce, G., Heeren, J.D., and Hill, T., "Orion SysML Model, Digital Twin, and Lessons Learned for Artemis I," *INCOSE International Symposium* 33 (2023): 290, doi:10.1002/iis2.13022.

12.16.   Ye, Y., Yang, Q., Zheng, J., Meng, S., and Wang, J., "A Dynamic Data Driven Reliability Prognosis Method for Structural Digital Twin and Experimental Validation," *Reliability Engineering & System Safety* 240 (2023): 109543, https://doi.org/10.1016/j.ress.2023.109543.

Autonomy can enhance operational safety and efficiency. Some systems use a hierarchical safety management strategy, categorizing safety thresholds into at least two levels based on the severity and impact of potential risks. When thresholds are crossed, specific safety modes correspond to each level and trigger different responses. This tiered approach allows for more nuanced responses to varying degrees of system health issues. At its foundation is a comprehensive mapping database that associates different safety thresholds with corresponding system safety modes. The database allows the system to quickly identify the current risk level by monitoring key parameters in real time and determine the most suitable safety mode to switch to. The database continuously updates with new data and insights gained from ongoing operations and system performance analysis.

Upon detecting that key operational parameters have reached or exceeded preset safety thresholds, the system autonomously starts a sequence of adjustments [13.1]. These adjustments can range from minor changes to system settings to more significant reconfigurations of spacecraft functionalities. The goal is to mitigate the detected issue while minimizing disruption to the mission and ensuring the continued safety and performance of the spacecraft. A system was designed to manage and resolve spacecraft issues using algorithms that consider various factors. These factors include the malfunction, current mission phase, and resources. The system determines the optimal response by downgrading or reconstructing spacecraft functionalities. This capability is crucial for maintaining mission continuity and maximizing the spacecraft's operational lifespan.

This structured approach to managing spacecraft safety through autonomous health management systems offers several benefits:

- Enhanced safety: By ensuring that the spacecraft can maintain critical functions even in the face of faults, the system enhances the overall safety of the mission.

- Reduced ground dependency: The ability to manage faults reduces the spacecraft's need for real-time ground intervention, which is crucial for missions beyond immediate Earth orbit or those with limited communication windows.

- Efficient resource use: By prioritizing essential functions and systematically managing the spacecraft's operational state, the system ensures efficient use of resources under constrained conditions.

This chapter will discuss autonomy, and how it can manage safety thresholds and adjust spacecraft functionalities to enhance operational safety, reliability, and mission success.

13.1.
What is Autonomy?

The concept of autonomy in spacecraft operations has evolved, becoming a cornerstone of modern satellite and transfer vehicle functionalities (Figure 13.1). One example is the Automated Transfer Vehicles (ATVs) and their autonomous docking capabilities. The evolution is driven by the need for spacecraft to function independently during periods without ground communication. This ensures the continuity of mission objectives and the ability to enter a stable safe mode when failures occur. The sophistication of autonomy in space missions encompasses a wide range of functionalities, from basic automatic operations to advanced autonomous decision-making processes that react dynamically to anomalies.

A combination of automatic and autonomous functions defines autonomy in spacecraft systems. Other concepts work together to create a robust and intelligent system capable of conducting space missions with minimal human oversight:

- Automatic functions are operations that the spacecraft executes according to a predefined schedule or through direct commands sent from mission control. These functions do not require the spacecraft to decide in real time. Examples include routine data collection, scheduled system checks, and prepro-grammed maneuvers. Automatic functions ensure the spacecraft performs essential tasks consistently and reliably, even when out of communication range with Earth.

- Autonomous functions elevate the spacecraft's operational capabilities by enabling it to decide and respond to anomalies during the execution of automatic functions. This includes diagnosing and addressing system failures, optimizing mission parameters in response to environmental changes, and deciding the best course of action in emergencies. Autonomous functions enhance operational flexibility and safety, allowing the spacecraft to protect itself and its mission in the dynamic space environment.

Figure 13.1 Autonomy terminology diagram.

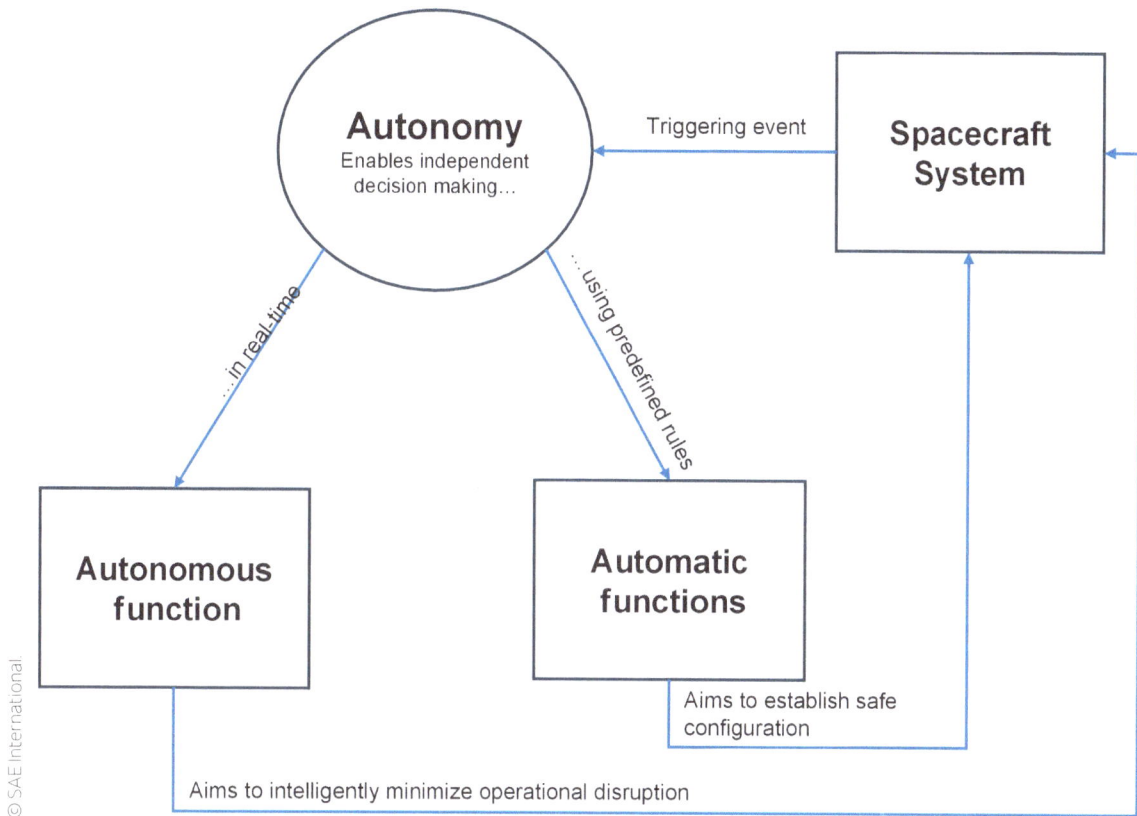

- Autonomy events are a term used to describe any anomaly or significant status change that triggers the spacecraft's autonomous systems. These events can range from system malfunctions and unexpected environmental conditions to deviations from the mission plan. When an autonomy event occurs, the spacecraft's autonomous functions activate to adapt or respond, ensuring that they maintain mission integrity. This capability is vital for long-duration missions where immediate ground intervention is not possible.

Spacecraft can have onboard intelligence that enables them to operate independently from ground support [13.2]. Designers equip spacecraft with advanced algorithms and computing capabilities to analyze data, decide, and execute operations autonomously. They can also adjust their mission strategies in real time, respond to unexpected challenges, and even learn from their environment to improve performance. This level of independence is important for missions beyond near-Earth space, where communication delays can be significant. For example, China's manned space missions have advanced the field of autonomous in-orbit health management, ensuring safety and reliability for long-term missions [13.3]. The concept of autonomy can encompass the entire ecosystem of a space mission, minimizing human intervention by intelligently distributing autonomy functions between the spacecraft and the ground segment. This system approach increases operational efficiency by enabling the spacecraft to handle complex tasks independently, while keeping the ground segment informed and in control of critical decisions. It represents a shift toward more dynamic and responsive mission operations.

13.1.1.
Levels of Autonomy

The ECSS plays a pivotal role in harmonizing the operations and technologies used in space missions. One of its significant contributions is the establishment of a structured framework to classify the levels of autonomy in spacecraft operations. This framework provides a common understanding and approach to implementing autonomy, ensuring consistency across various missions and space agencies. We divide the classifications into three major categories: mission execution autonomy (E-Levels), data management autonomy (D-Levels), and onboard fault management (FDIR).

Let us simplify these for better comprehension:

- Data management autonomy (D-Levels): This category focuses on how spacecraft manage and store mission data, particularly in response to communication outages or system failures:

  - D1 (low level): The spacecraft stores only essential mission data following a ground outage or a failure situation. This includes storing event reports and managing data storage efficiently.

  - D2 (high level): At this level, the spacecraft stores all mission data onboard, making it independent of ground segment availability. This comprehensive data storage

and retrieval capability ensures that the spacecraft does not lose any critical data, even without immediate ground support.

- Mission execution autonomy (E-Levels): This category defines a spacecraft's ability to execute its mission, ranging from reliance on ground control to full independence in decision-making and operations execution.

  - E1 (low level): At this level, the spacecraft primarily operates under direct control from the ground. It has limited onboard capabilities to address safety issues, relying on real-time commands from mission control for nominal operations and executing preprogrammed commands for safety-critical functions.

  - E2: Here, the spacecraft can execute preplanned, ground-defined mission operations autonomously. While real-time control from the ground is still necessary for nominal operations, the spacecraft can perform certain tasks independently, using time-tagged commands for safety issues.

  - E3: At this intermediate level, the spacecraft can execute adaptive mission operations onboard. This includes autonomously managing events and executing onboard operations control procedures with no need for real-time ground intervention.

  - E4 (high level): The highest level of autonomy, where the spacecraft can conduct goal-oriented mission operations entirely on its own. This involves replanning missions based on goals and executing complex operations independently, showcasing the spacecraft's advanced decision-making capabilities.

- FDIR autonomy levels: FDIR autonomy levels detail how spacecraft manage and recover from onboard failures:

- F1 (low level): The focus is on establishing a safe spacecraft configuration following an onboard failure. The crew is responsible for identifying anomalies, isolating failed equipment or functions, and placing the spacecraft in a safe state until they receive further instructions from the ground.

- F2 (high level): This level aims to reestablish nominal mission operations following an onboard failure. Beyond isolating failed components, the spacecraft reconfigures itself to resume normal operations and continue its mission objectives, showcasing a higher degree of self-reliance and resilience.

With these autonomy levels, the ECSS provided a standardized framework that can guide the development and implementation of autonomous systems in space missions. This ensures that spacecraft have the capabilities to manage their operations, respond to failures, and achieve mission objectives with varying degrees of independence from ground control.

13.2.
Enabling Technology vs. Process Improvement Technology

Autonomy serves two primary concepts. First, it acts as an "enabling technology" that allows systems to survive and operate without direct contact. Second, it serves as a "process improvement technology" that streamlines or reduces the cost of operation. For instance, enabling technologies are important for interplanetary missions, rovers, and military satellites, where independence from ground communication is vital. Process improvement technologies benefit missions like EO,

telecommunications, and navigation by optimizing control center operations and focusing data processing on user demands.

13.2.1.
As an Enabling Technology

This can include spacecraft systems, such as space probes, landers, rovers, and transfer vehicles. These systems need to make autonomous decisions whenever any direct or continuous communication with Earth is not feasible. We can characterize them with:

- Sophisticated onboard command sequences: Ranging from executing macro command sequences to using intelligent onboard mission planners, these systems show advanced operational capabilities. Technical specifications: The designers have designed the OBSW architecture with modularity, incorporating real OSs and multi-CPU board architectures. This includes separate OBCs for payloads and ensures higher SW layers are HW-independent.

- Verification and validation: Maintaining and verifying these complex autonomous systems requires sophisticated spacecraft test benches. These test benches simulate detailed scenarios, including error injection mechanisms, to test the spacecraft's error identification capabilities across various simulated conditions. Specifically, we rigorously test the autonomous functions of OBSW using scenario verification facilities (SVF) to ensure reliability and functionality.

13.2.2.
As a Process Improvement Technology

This can include systems involved with mission product generation like EO satellites to streamline operations. Key features include:

- User-driven mission control: Unlike traditional methods where customers specify operational details, the modern approach allows them to define observation targets, desired product characteristics, and delivery timelines. This customer-focused model greatly improves our ability to adapt and respond to our mission.

- Intelligent mission planning: By integrating archive data with real-time observations, an intelligent mission planning system autonomously generates command sequences. This system ensures efficient observation, data downlink, and the amalgamation of new and archived data to fulfill customer requests.

- Semiautomatic operations: This autonomy level facilitates semi-automatic operations, allowing for single-shift management of spacecraft and payload control, optimizing resource use and operational costs.

To validate the benefits, detailed scenario simulations are essential that will not only encompass spacecraft operations but also include ground segment functionalities, such as user-request-based mission planning. Such comprehensive testing ensures the seamless integration of spacecraft and ground operations, highlighting the system's efficiency and effectiveness in meeting mission objectives. In addition, recent advancements in ML can help speed up testing for:

- New environments: Algorithms can analyze data from the spacecraft's sensors to adapt to new environmental conditions, such as changes in solar radiation or unexpected obstacles on a planetary surface.

- Predicting system failures: By analyzing patterns in the spacecraft's operational data, AI can predict potential system failures before they occur, allowing for preemptive maintenance or adjustments to avoid critical failures.

- Optimizing mission strategies: Simulating various mission scenarios and strategies, allowing spacecraft to select the most efficient course of action in real time, maximizing mission success and scientific return.

Integrating autonomy for health management represents a shift toward more resilient and efficient space missions. It offers valuable directions for effectively predicting equipment behavior and optimizing operational strategies in space missions in spacecraft health management research. However, the direct application of many studies to the space domain may encounter unique challenges, such as adapting models to the sparse and highly variable data environment of space operations. A novel framework proposed by Calabrese et al. for PHM in industries combines edge-cloud infrastructure with ML for real-time analysis [13.4]. This semisupervised, partially online approach addresses the scarcity of historical data, a common hurdle in spacecraft health management. While promising, the scalability and adaptability of this framework to the constrained computational resources available on spacecraft need further investigation. Das et al. proposed a DT framework to help understand a machine intelligence approach for anomaly detection and health status prediction [13.5]. Again, these methodologies show potential applications by offering system-wide monitoring capabilities. However, the complexity of space missions causes a more robust validation of these models against the extreme conditions encountered in space.

13.3.
Requirements for Autonomous Health Management

Existing discussions on integrating autonomy within health management systems in spacecraft reveal a significant shift toward enhancing in-orbit independence and safety, reducing the reliance on tracking, telemetry, and command resources [13.6]. Based on the discussion in this chapter, we can summarize the requirements for such systems as follows:

- Comprehensive health status monitoring: The system must continuously collect and analyze the health status of all critical spacecraft subsystems, including energy, attitude control, thermal management, and life support systems. This is a typical requirement for any data analysis, fault detection, and decision-making task, along with data acquisition devices to collect status data from various spacecraft components.

- Real-time fault detection and diagnosis: The system must be capable of identifying and diagnosing faults in real time, using algorithms and models that can accurately interpret sensor data and system states. This is an essential feature for making autonomous decisions regarding fault isolation, system reconfiguration, and recovery actions without waiting for ground commands, especially in scenarios where immediate action is critical.

- Layered fault management and recovery: A by-level safety threshold management approach may be required to handle progressive faults. This will ensure that the health management system can maintain essential functions and enter a safe mode if necessary.

- Distributed processing architecture: Because of the complexity and distributed nature of spacecraft systems, the health management functionalities should distribute across multiple computers within different subsystems. Each subsystem will be responsible for specific tasks. The overall design will become more complex, so it is important to include mechanisms for testing and verifying these functionalities on the ground and in orbit to ensure reliability and effectiveness.

- Scalability and adaptability: The system should be scalable to accommodate additional modules or changes in spacecraft configuration and adaptable to new mission requirements or operational scenarios.

- User interface: The user interface should deliver timely information to both astronauts and ground personnel regarding the spacecraft's health, fault diagnostics, and recovery actions.

While autonomous health management systems have promising capabilities for enhancing spacecraft safety and efficiency, they also pose challenges in terms of computational capability, design complexity, and the need for thorough verification and validation [13.7]. The development of detailed simulations, or test benches capable of error injection and modeling complex error symptoms, is essential for testing the OBSW's error identification mechanisms. For example, the symbiotic system-of-systems design framework has shown promising results for safe and resilient autonomous robotics, with applications in both space and industrial domains [13.8]. As spacecraft operations evolve toward higher levels of autonomy, integrating intelligent functions for autonomy between the space segment and ground segment is essential.

13.3.1.
System Architecture

Considering the complexity and distributed nature of spacecraft systems, we can structure autonomous health management across multiple levels to enhance efficiency and reduce the computational load on the central processor.

- System level (core processing unit): This level oversees the entire health management operation, deciding based on comprehensive data analysis.

- Subsystem level (control computers or regional controllers): Each subsystem of the spacecraft, such as life support or propulsion, has its controller that manages the health of its specific domain, reporting to the core processor.

- Equipment level (subsystem internal subnet equipment and terminals): At the most detailed level, individual parts and terminals carry out the assigned health management tasks given by their respective subsystem controllers.

Figure 13.2 illustrates the multitiered approach for rapid failure response and distributed processing across the spacecraft's network. It also enables each subsystem to manage its health autonomously to some extent. This architecture not only enhances the spacecraft's overall safety and reliability but also significantly reduces the demand for ground-based tracking and control resources, making missions more efficient and cost-effective. By leveraging such a by-level health management system, spacecraft can maintain optimal operation through autonomous monitoring, fault detection, and recovery processes, ensuring mission success even in the most challenging environments. In the event of a major emergency failure, the system can place the spacecraft into a safe mode, isolating the

issue until ground-based teams can investigate and resolve the problem. For less critical failures, the system can issue alarms and can attempt to rectify issues through actions such as system resets or component shutdowns. This architecture also provides functionality across different levels of spacecraft operation:

- Communication with crew and ground control: The communication system has a key feature that allows for the transmission of vital health status information, fault diagnoses, and recovery actions to both the onboard crew and ground control teams. This ensures that all stakeholders, including the onboard crew and ground control teams, receive real-time updates on the spacecraft's condition, enabling coordinated response efforts when necessary.

- Hierarchical fault detection and diagnosis: An approach that enhances the ability to identify and address system anomalies quickly in fault detection and diagnosis.

  - Equipment level: At a detailed level, we can use built-in test capabilities, HW monitoring circuits, and test points to check the status of individual devices. Devices that are directly connected to the spacecraft's main system network send fault information directly to the core processing unit for immediate action.

  - Subsystem level: Each subsystem controller monitors the health of its respective domain, executing fault detection and diagnosis based on predefined fault modes and diagnostic programs. The subsystem controllers report detected faults to the core processing unit. If a subsystem controller identifies a condition that requires entering security mode, it can start emergency procedures directly or defer to the core processing unit for decision-making.

Figure 13.2 Hierarchical model for autonomous health management.

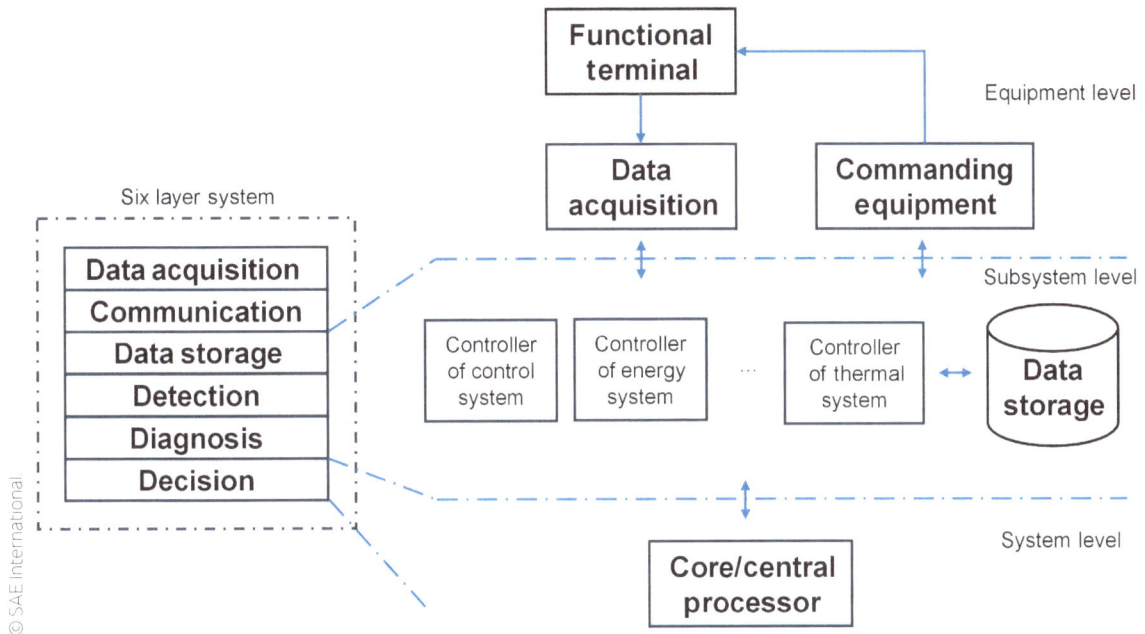

- System level: The core processing unit aggregates fault information from across the spacecraft's subsystems to assess whether conditions warrant entering safe mode. It then coordinates the execution of emergency protocols as needed. The unit communicates all fault information, along with instructions for response, to both the crew and ground control.

- Emergency handling and information dissemination: In the event of significant system failures, the architecture enables transition into a safe operational state, minimizing the risk of further damage or mission compromise. This response is critical for maintaining the safety of both the crew and the spacecraft, especially during periods when direct intervention from ground control is not feasible.

Collecting and analyzing health status data from every level of spacecraft operation ensures that we identify and address potential issues promptly. This layered approach to health management, from individual components to the spacecraft, demonstrates the sophistication and necessity of autonomous systems in modern space missions.

13.3.2.
Establishing Safety Thresholds

The concept involves creating a comprehensive mapping database that correlates various safety thresholds of critical risks with specific system safety modes. As spacecraft encounter different operational scenarios, we monitor critical parameters against preset thresholds. When the system reaches these thresholds, it starts a cascading series of actions to mitigate risks, including sending alarms, shutting down high-power devices, and deactivating payloads. This structured degradation and reconstruction of spacecraft functionalities ensures optimal autonomous response, bridging the gap between the unpredictability of automated processes and the need for continuous mission operation.

The initial step involves pinpointing safety-related fault modes. This analysis is critical for understanding the potential risks and preparing the system for autonomous management. After this, it is important to categorize different fault modes and assign system safety modes postdisposal. The aim is to streamline the process, enabling a single disposal solution to address multiple fault scenarios efficiently. For instance, both reducing power supply capacity and diminishing heat dissipation capability might trigger the same response—shutting down specific devices. This consolidation simplifies implementation and bolsters system reliability. The system can dynamically control the spacecraft functionalities according to the severity of detected faults, aligning them with corresponding safety modes. This stepwise degradation process culminates in a minimum safety mode, ensuring the spacecraft keeps essential operational capabilities. The minimum safety mode encompasses critical functions such as downlinking engineering telemetry data, receiving ground commands, importing data, maintaining a stable thermal environment, and controlling attitude.

13.3.3.
Downgrading to Minimum Capability

This procedure ensures that, in the event of a fault, the spacecraft can still maintain essential functions to preserve its integrity and safety. The process allows for logical structuring and execution in stages, enabling a controlled

response to various fault conditions [13.4]. This is a detailed explanation of the procedure:

- Procedure design: The team programs the spacecraft with a set of procedures that correspond to different safety modes. These procedures prioritize essential functionalities and systematically reduce the spacecraft's operational capacity as needed to respond to faults. The design ensures that critical operations can continue, albeit at a reduced capacity.

- Fault occurrence response: Upon detecting a fault, the system assesses which safety mode is relevant based on the fault. The spacecraft then disables the functionalities associated with that safety mode, effectively downgrading to a lower level of operational capability. The spacecraft keeps reducing its functionalities step by step until it reaches the minimum safety mode.

- Automatic functional degradation: The spacecraft autonomously manages this process, deciding based on predefined thresholds for various faults. As each threshold is reached or a fault is detected, the spacecraft autonomously follows a predetermined sequence of degradations. It transitions from higher safety modes to the minimum safety mode. Throughout this process, the spacecraft continuously collects data to monitor the status of the fault.

- Real-time data collection: During the degradation process, the spacecraft collects data in real time to assess if it has resolved the fault. If the fault is cleared, the system stops further degradation processes and stabilizes the spacecraft in the current safety mode. This dynamic assessment allows for responsive adjustments to the spacecraft's operational state.

- Ground-side diagnosis and treatment: Once the spacecraft is in a stable safety mode and reenters a zone where ground monitoring and control are possible, ground personnel undertake a detailed analysis. They use the data stored on the spacecraft, along with real-time telemetry, to diagnose the fault. Based on this analysis, ground teams can then implement corrective actions to resolve the fault and potentially restore the spacecraft to its nominal operational state.

13.4.
Demonstrations of Autonomy

13.4.1.
Project for Onboard Autonomy (PROBA) Satellite

The PROBA satellites, launched in October 2001, are part of the ESA's initiative to test and validate advanced technologies in orbit (Figure 13.3). They focus on enhancing onboard autonomy and automation levels in satellite operations. These satellites show the feasibility and benefits of incorporating high levels of autonomy into spacecraft systems, aiming to reduce the reliance on ground control and improve operational efficiency and flexibility. It was notable for several pioneering achievements in onboard autonomy:

- Thirty-two-bit Space application microprocessor: It was the first ESA satellite to use Europe's 32-bit space application microprocessor, the ERC32 chipset, marking a significant advancement in onboard computing power.
- DSP: PROBA-1 employed a DSP as an instrument control computer (ICU), enhancing the satellite's data processing capabilities.
- Autonomous star sensor: The mission featured the first in-orbit application of a newly designed autonomous star sensor for attitude determination, enabling the satellite

to identify star constellations without ground intervention.
- Onboard GPS: PROBA-1 was among the first ESA satellites to use onboard GPS for orbit determination, improving its navigation accuracy.
- OBSW innovations: The satellite's OBSW was notable for several firsts.
- It was ESA's first OBSW coded in C instead of Ada, a programming language traditionally used for space applications.
- It operated on an actual OS (VxWorks), moving away from purely Ada-coded OBSW implementations.
- The mission validated the GNU C compiler for the ERC32 target by successfully operating GNU C compiled OBSW on the ERC32.

Achievements and functionality: PROBA-1 demonstrated new onboard functionalities that significantly advanced the state of autonomy in space missions:

- GPS-based position determination: For the first time, an ESA satellite could determine its position in orbit using GPS, enhancing its navigation capabilities.
- Autonomous attitude determination: The active star sensor enabled autonomous attitude determination by automatically identifying star constellations.
- Autonomous navigation event prediction: The satellite could autonomously predict navigation events, such as target and station flyovers, facilitating better mission planning.
- Limited onboard mission planning: Based on its autonomous capabilities, PROBA-1 featured a limited onboard mission planning functionality, allowing for some degree of autonomous decision-making regarding its operations.

Figure 13.3 Illustrates the launch of ESA's PROBA satellite, which served as an in-orbit technology demonstration and evaluation platform. It aimed to assess new HW, SW, onboard operational autonomy, as well as EO and space environment monitoring instruments.

ESA

The technologies showed by PROBA-1 and its successors in the PROBA series have significant implications for spacecraft health management. The development of spacecraft capable of self-managing their health and operational status is aided by advancements in onboard autonomy, particularly in areas like autonomous navigation, attitude determination, and mission planning. These capabilities are crucial for long-duration missions, deep space exploration, and scenarios where immediate ground intervention is not feasible, enhancing the resilience and safety of space missions.

13.4.2.
Mars Science Laboratory (MSL)— Curiosity Rover

The rover represents significant progress in the application of autonomy and onboard health management systems during space exploration. Launched by NASA on November 26, 2011, Curiosity landed on Mars on August 6, 2013 (Figure 13.4). Its primary mission was to investigate Mars' Gale Crater as a potential habitat for microbial life, study the planet's climate and geology, and collect data for a future manned mission to Mars. Curiosity's design incorporates advanced autonomy features to navigate and conduct scientific experiments with minimal direct control from Earth. Given the communication delay between Earth and Mars, which can be up to 22 min one way, the autonomous capabilities were crucial for the rover's operational efficiency and safety.

Figure 13.4 The Mars Curiosity rover.

NASA/JPL-Caltech

- Autonomous fault protection: Curiosity has a sophisticated fault protection system that autonomously detects, isolates, and recovers from onboard anomalies. This system is crucial for managing the rover's health, especially given the communication delay between Mars and Earth.

- Automated data analysis: The rover can analyze data from its instruments and sensors independently to detect signs of potential issues. It can adjust its operations based on the findings, such as instrument degradation or unexpected environmental conditions.

- Self-monitoring capabilities: Curiosity constantly monitors the health of its power, thermal, and mechanical systems. It uses this information to make decisions that ensure its survival, like adjusting activities to manage power and thermal loads.

13.4.3.
LRO

LRO is a NASA mission launched on June 18, 2009, with the primary goal of mapping the moon's surface to support future lunar exploration missions (Figure 13.5). LRO's suite of instruments has provided crucial data on the lunar environment, including topography, surface composition, and potential water ice deposits. A key factor in LRO's success is its advanced autonomy and onboard health management systems [13.9]. These systems allow it to efficiently and reliably perform extended operations in lunar orbit. The health management features included:

Figure 13.5 The LRO. It carries seven instruments that make comprehensive remote sensing observations of the moon and measurements of the lunar radiation environment.

NASA/GSFC.

- FDIR: LRO's onboard systems automatically detect and respond to potential faults. This includes switching to backup systems in the event of an HW failure and entering a safe mode to await further instructions from Earth if necessary.

- Thermal and power management: The spacecraft autonomously manages its thermal control systems to keep its instruments and electronics within operational temperature ranges. It also optimizes power usage based on the available solar energy, ensuring that critical systems remain powered during lunar eclipses.

13.4.3.1.
Autonomous Spacecraft Management (ASM) Projects

ASM projects represent a significant shift toward more self-reliant, efficient, and robust space missions. These projects aim to reduce the reliance on ground-based control and increase the spacecraft's ability to decide, perform tasks, and manage its health autonomously. This shift is crucial for deep space missions, where communication delays with Earth can make real-time management impractical, and for enhancing the operational efficiency and safety of satellites in Earth orbit. Some capabilities include:

- Intelligent agents: Some ASM projects employ intelligent SW agents that autonomously manage spacecraft health by monitoring system performance, predicting failures, and implementing preemptive maintenance or reconfiguration strategies.

- ML is utilized in anomaly detection projects to analyze telemetry data, identify patterns

that signal system degradation or impending failures, and initiate corrective actions autonomously.

- Distributed health management: ASM approaches may distribute health management tasks across multiple spacecraft modules or systems, allowing for a more resilient and responsive approach to managing spacecraft health.

13.5.
Summary

This chapter focuses on the significance of autonomous health management systems in modern spacecraft operations. It draws discussions from several research articles and highlights advancements in architecture, fault detection, and management strategies.

Incorporating autonomy offers significant benefits, including:

- Enhanced diagnostic capabilities: Autonomous systems can perform continuous health checks, diagnose potential issues, and implement corrective actions without waiting for ground-based commands.

- Predictive maintenance: By analyzing data trends, autonomous systems can predict potential failures before they occur, allowing for preemptive maintenance and reducing the risk of critical failures.

- Operational flexibility: Autonomy enables spacecraft to adapt to changing mission conditions, optimizing performance and extending mission life.

While the proposed concepts and technologies show potential benefits in enhancing spacecraft safety and reducing ground management, the practical implementation and integration of these systems remain challenging. Autonomy in spacecraft systems is an enabling technology that plays a crucial role in advanced space missions. These missions are characterized by minimal ground contact or significant signal delays, such as those encountered in deep space exploration. The autonomy focuses primarily on the OBSW rather than the ground system, allowing spacecraft to independently perform critical functions and ensure mission success under challenging conditions.

These include ensuring system reliability, managing the complexity of distributed architectures, and adapting to the dynamic conditions of space missions.

A few implementations have shown high-level autonomy in action by managing multiple complex tasks without ground intervention, including:

- Automatically activating payloads based on the mission's evolving needs and objectives.
- Continuously assessing the spacecraft's health and implementing FDIR strategies to address any anomalies or malfunctions.

To achieve this level of autonomy, an architecture is needed. It should involve a central control module that closely integrates with subsystem controllers, parameter monitors, and an event manager. This integration ensures dynamic responses to operational challenges, thereby maintaining mission integrity and safety. Also, while autonomy as enabling technology is crucial for spacecraft operations, its role as a process improvement technology focuses on optimizing

tasks related to the spacecraft's ground segment. This includes enhancements in:

- Ground segment infrastructure: Streamlining operations to reduce the need for constant human intervention and enable more efficient mission control.
- Mission product handling: Automating the processing, storage, and distribution of mission data to improve accessibility and usability for end users.
- Mission timeline generation/optimization: Using advanced SW systems to automate the planning and optimization of mission timelines, ensuring optimal use of resources and mission opportunities.

The discussions highlight the innovative application of AI technologies in predictive maintenance and anomaly detection, underscoring their potential to revolutionize spacecraft operations. While many studies have collectively shown the potential of autonomy in advancing spacecraft health management, several critical areas require further investigation. These include adapting models to the unique data environment of space, considering the computational constraints of spacecraft systems, and validating these technologies against the harsh conditions of space missions. Future research should focus on the scalability of health management systems, integrating AI and ML for predictive maintenance, and developing standardized protocols for autonomous health management across missions. Real-world testing and validation of these systems in various space mission scenarios would provide valuable insights into their effectiveness and areas for improvement. As the field progresses, addressing these challenges will be crucial for realizing the full potential of autonomy in this domain.

References

13.1. Lei, Y., Qu, Q., Liang, D., Mao, Y. et al., "Spacecraft System Autonomous Health Management Design," in Li, X., Liu, C., Zhang, J., and Kim, S.I. (eds.), *Signal and Information Processing, Networking and Computers: Proceedings of the 5th International Conference on Signal and Information Processing, Networking and Computers (ICSINC)* (Singapore: Springer, 2019), 310-316.

13.2. Jayaram, S., Johnson, R.W., and Prasad, G., "Near Real-Time Autonomous Health Monitoring of Actuators: Fault Detection and Reconfiguration," in *Florida Conference on the Recent Advances in Robotics*, Gainesville, 2004.

13.3. Bai, L., Yang, H., and Liang, K., "The Research of Autonomous In-Orbit Health Management System for China's Manned Spacecraft," *Matec Web of Conferences* 257 (2019): 01006.

13.4. Calabrese, F., Regattieri, A., Bortolini, M., Gamberi, M. et al., "Predictive Maintenance: A Novel Framework for a Data-Driven, Semi-Supervised, and Partially Online Prognostic Health Management Application in Industries," *Applied Sciences* 11, no. 8 (2021): 3380.

13.5. Das, S.N., Ahuja, M., Singi, K., Dey, K. et al., "Digital Twin Based Fault Analysis in Hybrid-Cloud Applications," In *Proceedings of the 10th IEEE/ACM International Workshop on Software Engineering for Systems-of-Systems and Software Ecosystems* pp. 29-32.

13.6. Liang, K., Deng, K., and Ding, R., "Autonomous On-Orbit Health Management Architecture and Key Technologies for Manned Spacecraft," *Manned Spacefl* 20 (2014): 116-121.

13.7. Doyle, R., Kubota, T., Picard, M., Sommer, B. et al., "Recent Research and Development Activities on Space Robotics and AI," *Advanced Robotics* 35, no. 21-22 (2021): 1244-1264.

13.8. Mitchell, D., Blanche, J., Zaki, O., Roe, J. et al., "Symbiotic System of Systems Design for Safe and Resilient Autonomous Robotics in Offshore Wind Farms," *IEEE Access* 9 (2021): 141421-141452.

13.9. Eaty, N.D.K.M. and Bagade, P., "Digital Twin for Electric Vehicle Battery Management with Incremental Learning," *Expert Systems with Applications* 229, no. Part A (2023): 120444, doi:https://doi.org/10.1016/j.eswa.2023.120444.

This book provided some insights into health management and revealed how AI and autonomy are revolutionizing the field. For the space domain, a lot of these developments are being promoted by New Space—a term coined for the recent wave of private-sector involvement in space exploration and development. The motivation behind these New Space companies varies, with some seeking commercial profit and others aiming to advance scientific research and increase accessibility and affordability in the space sector. This has also opened possibilities for new business models that were simply not feasible with the traditional government-led space programs. Private companies now provide alternative services and solutions to others rather than owning whole assets. In a competitive environment with high service demands, a strong health management system is crucial for monitoring and analyzing spacecraft health and performance in real time. This helps find potential problems, anticipate failures, and take preventive maintenance actions. Even if anyone is in the market to hire or lease a spacecraft, they would want to know about its condition.

Perhaps, the unique and challenging nature of space exploration engineering has progressed to a point where health management technologies are not just beneficial but have become quintessential.

One-off product: Traditional statistical quality control methods are less applicable to space missions because each spacecraft is often a unique, one-off product. The bespoke nature of these missions shows that relying solely on large datasets from identical units cannot ensure reliability and performance.

Nonrepairability: Since designers assume that spacecraft cannot be repaired once launched, they shift their focus toward designing systems that can anticipate, detect, and adapt to failures, and independently take actions in response to them. The criticality of FDIR features highlights the importance of preemptive identification of potential issues and addressing them without human intervention.

- The vast distances in space exploration result in autonomous systems capable of rapid response to unforeseen conditions or failures. Self-assessment of health and reconfiguration is necessary, without waiting for delayed commands from ground control.

- The unpredictable nature of space missions necessitates systems that are both reconfigurable and adaptable. Since the full scope of problems may only emerge upon arrival at the destination, having a system that can adapt to new challenges and reconfigure itself as needed is essential. It is important to guarantee that these systems can predict and adjust to changes in their operating environment or mission parameters.

- Transparency in the process: Essential for mission planning and ensuring the resolution of anomalies, process transparency guarantees the ability to reproduce spacecraft behavior on the ground in a deterministic manner.

- Many health management solutions enhance transparency by providing detailed behavior and performance data. This ensures that ground teams have the necessary information to understand and replicate all scenarios, facilitating better decision-making and problem-solving.

To fully unlock this potential, several notable technologies and techniques have emerged. These include DL algorithms for predictive analysis, advanced sensors and diagnostics, robotics, autonomous decision-making systems, edge computing architectures, and adaptive reconfiguration capabilities. The rest of this chapter explores how each technology contributes to health management systems' goals. I will share my perspective on their impact, challenges, and potential future developments. This exploration will not only emphasize the critical role of these technologies in enhancing space mission success, but also reflect on their broader implications for the field.

14.1.
AI-based Systems in Space

Incorporating AI in space has already revolutionized how we approach aerospace maintenance and repair. AI-based systems process data quickly, enabling real-time monitoring, diagnosis, and proactive issue identification. This ability is crucial in preventing serious failures that could terminate or jeopardize missions. As a quickly progressing field with significant contributions from various research efforts, here are some developments based on recent studies:

- The application of DL models for predictive maintenance in the industrial sector showcases the potential of AI in detecting and predicting failures before they occur. The utilization of GA for optimizing the performance of Bi-LSTM models is noteworthy, as it addresses one of the significant challenges in AI: the selection of optimal hyperparameters. However, additional validation may be necessary for the direct applicability of these findings to spacecraft systems because of the unique operational and environmental conditions encountered in space.

- By fine-tuning AI models for predictive maintenance in space's harsh conditions, we can improve spacecraft reliability and durability, potentially reducing costs and risks in space missions.

- SpaceDrones 2.0: Some researchers have examined the use of computer vision and AI in autonomous space-borne applications and propose a novel approach for simulating and validating space-based computer vision. Such studies have emphasized the role of synthetic training and domain randomization in preparing computer vision models for the unpredictable conditions of space.

  - A major obstacle in space exploration is the lack of extensive real-world data for training AI models. However, the use of synthetic data and simulations also prompts questions about the models' ability to generalize to actual space conditions.

  - There, the development of robust, synthetic training methods could aid in implementing AI in these applications, enabling more autonomous and flexible missions. By simulating and validating AI models on Earth before deployment, we could enhance the success rates of missions, particularly in deep space exploration, where human intervention is restricted.

- Several studies have investigated the implementation of diagnostic and prognostic tools for industrial machines, using vibration monitoring and novelty detection to infer wear from critical components. Most of these principles, from CBM and the use of AI for anomaly detection, apply to spacecraft health management. Adapting these strategies to spacecraft systems comes with additional complexities. These include operating in a vacuum and being exposed to radiation, which may impact the performance and reliability of AI models.

14.1.1.
Explainable AI (XAI)

Explainability in AI systems is crucial for gaining insight into the decisions made by these technologies. This is especially true in spacecraft health management, where decisions can have significant outcomes. XAI acts as a bridge, enabling AI's complex algorithms to be understandable and trustworthy for human operators. Trust is essential in the space sector, where the environment is unforgiving and the risks are high. XAI not only enhances transparency in AI decisions, but also builds confidence among engineers and operators. This trust improves the safety and efficiency of space missions.

Regulatory bodies extensively oversee the space industry to ensure mission safety and reliability. XAI serves a vital role in meeting these regulatory standards by making AI-driven decisions clear and understandable, thus aligning spacecraft health management systems with global safety norms. XAI is not just about building trust and complying with regulations. Through the process of making AI decisions explainable, engineers can identify and correct biases or errors, enhancing the accuracy and adaptability of AI algorithms.

With a forward-thinking mindset, XAI will play a central role in the future of space exploration. Its integration into spacecraft health management holds the potential to enhance operational transparency and redefine human–machine collaboration. Table 14.1 provides a simplified breakdown that emphasizes the transformative potential of XAI in spacecraft health management. It showcases how XAI enhances mission success and paves the way for future advancements in space exploration.

Table 14.1 The importance of explainability of AI for safety-critical applications.

Aspect	Importance of XAI	Impact on missions
Understanding AI decisions	Essential for illustrating AI logic.	Improves trust and operational efficiency.
Trust and transparency	Builds confidence in AI systems.	This results in safer and more reliable missions.
Regulatory compliance	Ensures AI decisions meet safety norms.	Aligns with global standards for space exploration.
Continuous improvement	Highlights areas for AI model refinement.	Increases accuracy and adaptability of AI systems.
Future vision	Central to advancing space exploration.	Promises enhanced transparency and collaboration.

© SAE International

14.1.2.
Lack of Training Data

A significant obstacle to achieving autonomous spacecraft with ML lies in the absence of comprehensive training datasets that encompass different fault scenarios. To train ML algorithms in autonomous systems, large and diverse datasets are essential for recognizing, predicting, and responding to different operational scenarios, such as faults. However, there are several factors that make the creation of datasets for spacecraft particularly challenging:

- Rarity of faults: Space missions are designed to be highly reliable, resulting in a scarcity of fault data. This lack of real-world fault occurrences makes it difficult to develop robust ML models that can predict or identify faults accurately.

- Cost and risk of fault simulation: Simulating faults in spacecraft systems can be extremely costly and risky. Full-scale testing or introducing faults deliberately in operational systems is often not feasible.

- Complexity and variability: Spacecraft systems are complex, and the types of faults that can occur are highly variable and often unique to specific missions or environments.

Researchers and engineers are currently engaged in finding multiple solutions to overcome these difficulties.

- Synthetic data generation: To overcome the lack of real fault data, synthetic datasets are generated using simulations and DTs. For example, the SPEED+ dataset combines synthetic images and hardware-in-the-loop images from the Testbed for Rendezvous and Optical Navigation (TRON) facility to address domain gap challenges in spacecraft pose estimation.

- Transfer learning: Transfer learning techniques are used to build robust classification models with limited fault data. These models are initially trained on large, general datasets and then fine-tuned with the specific data available from spacecraft systems. This approach helps in leveraging existing knowledge to improve model accuracy in fault detection and classification.

- Federated learning: Decentralized federated learning methods, such as the Byzantine-fault-tolerance decentralized federated learning (BDFL) method, are employed to improve the robustness of ML models. These methods allow for training models across multiple decentralized datasets without sharing sensitive data, thus preserving privacy while enhancing performance.

- Onboard AI/ML implementation: Spacecraft like Operations Satellite (OPS-SAT) have demonstrated the use of onboard AI for autonomous operations, including image classification and unsupervised learning for fault detection and recovery. This practical implementation helps in validating AI/ML approaches in real mission scenarios.

- Enhanced datasets: Efforts are being made to develop new datasets specifically tailored for spacecraft fault detection and autonomous navigation. The SPEED-UE-Cube dataset, for instance, aims to facilitate the advancement of ML algorithms for vision-based navigation of noncooperative targets in spacecraft.

14.2.
Robotic Autonomy

Another progress is the ability of robots to function independently, which brings a significant change in how we maintain and service space. Engineers can design autonomous robots for complex maintenance tasks in inhospitable space environments, which humans cannot access due to extreme conditions such as vacuum, extreme temperatures, and high radiation levels. Their capacity to carry out repairs and adjustments significantly extends the operational lifespan of spacecraft, marking a critical development in space technology.

The combination of robot autonomy and AI-driven systems has given rise to the concept of in-orbit servicing. This involves performing maintenance, repairs, refueling, or upgrades on satellites and spacecraft while they are in orbit. Previously, many regarded these ideas as impractical or not cost-effective. This shows a transformative impact that has opened new possibilities by reducing the need for direct human intervention and overcoming the limitations imposed by human vulnerabilities to space conditions.

The capability of these robots to conduct servicing during missions has significant implications for space operations. It helps minimize waste and maximize the use of existing assets, contributing to sustainability and scalability. It could pave the way for the development of new business models and economic opportunities in space, such as satellite servicing companies or debris removal services.

Table 14.2 summarizes the discussion of how AI and robotic autonomy contribute to mission resilience, operational efficiency, commercial space operations, and the future of space exploration and commercialization. It integrates insights from scholarly articles to provide a comprehensive overview.

Table 14.2 Advantages of AI and robotic autonomy in space mission.

Aspect	Key insights	Integration and application
Mission resilience	AI-driven autonomy enables spacecraft to analyze challenges instantly and respond to them, minimizing mission failure risk.	Applying AI for optimal energy management in space activity can boost resilience and extend mission durations.
Operational efficiency	AI and robotics facilitate predictive maintenance and automated problem resolution, enhancing spacecraft operational periods.	We can adapt life cycle learning from symbiotic systems for space missions, promoting continuous adaptation and efficiency.
Commercial space operations	Efficient health management systems reduce downtime and extend satellite operational life, optimizing costs.	AI planning frameworks can improve the autonomy of space missions, allowing for adaptive responses to improve commercial viability.
Space exploration and commercialization	AI and robotic autonomy could cause autonomous missions with minimal human oversight, opening new frontiers in exploration.	Integrating intelligent systems in space missions supports autonomous exploration and operational efficiency, promoting space commercialization.

14.3.
Edge Computing and Onboard Data Processing

When carrying out the research for this book, I noticed an overwhelming number of publications on the topic of edge computing. This technology assures to change how we can carry out data processing in a more power-efficient manner and facilitate on-the-spot data processing directly on a spacecraft. It can decrease the need for ground-based decision-making, making it a worthwhile investment for designing deep space missions, where there are substantial delays during real-time communication with Earth. Table 14.3 summarizes some of the notable challenges.

TABLE 14.3 The challenges of edge computing for the field.

Challenges	Description	Strategies for overcoming
Local data processing	Processing data locally at the edge can result in latency and resource allocation challenges.	Implement lightweight algorithms, optimize data processing workflows, and use efficient data management methods.
Improved data security	Maintaining data security at the edge is complex because of the distributed nature of edge devices.	Employ robust encryption, secure data transmission protocols, and regular security updates and patches.
Enhanced mission resilience	Ensuring operational continuity in the face of failures or attacks is crucial for edge computing.	Optimize ML models for edge deployment, use model quantization, and leverage edge AI chips that are tailored for edge computing.
ML payloads	Running ML models at the edge requires finding a balance between computer power and efficiency.	Optimize ML models for edge deployment, use model quantization, and leverage edge AI chips tailored for specific purposes.
Efficiency in data utilization	Maximizing the utility of data managed at the edge to drive decisions.	Implement data analytics and processing techniques that reduce redundancy and prioritize valuable insights.
Power requirement of AI accelerators onboard	AI accelerators at the edge increase power consumption, affecting device longevity.	Opting for energy-efficient AI accelerators, implement power management strategies, and use solar power backups.

© SAE International

14.4.
DTs

A DT is more than just a standalone spacecraft simulator. It can automatically ingest configuration data from concept design tools, simulate performance in a virtual environment, and compare output data with system definition models. The digital fluidity facilitates a rapid way to verify spacecraft designs as they are being conceived by exchanging data between models of estimation, simulation, and definition.

There is a lot of interest in the application of DTs for the predictive maintenance of spacecraft by simulating various components and predicting failures before they occur. Through developing a virtual replica, engineers can monitor system health, simulate scenarios, and assess spacecraft responses to anomalies or environmental changes in real time. DTs also enable the optimization of the design with no expensive physical prototypes and still provide a realistic and dynamic environment to simulate missions and prepare for unexpected situations.

One of the main challenges they face is integrating and managing large volumes of data from different sources, such as sensors, operation logs, and environmental data, to ensure the accuracy of the DT. There are just not enough examples or case studies on the topic. I would assert that the complexity of simulating a spacecraft in real time would require significant computational resources, which is a barrier to analyzing more complex missions. As a result, many simulations might be overly simplified, resulting in incorrect predictions or assessments, potentially compromising the mission. **Table 14.4** summarizes the other issues.

Table 14.4 The challenges of understanding accurate DTs.

Challenge	Description	Strategies for overcoming
Data limitations	Space missions produce distinct datasets that are frequently restricted in size, making the training of resilient ML models challenging.	Developing synthetic data generation techniques and transfer learning approaches can alleviate the scarcity of space-specific datasets.
Real-time decision-making	The need for real-time decision-making in deep space missions pushes the limits of AI systems' current capabilities, which may require significant computational resources.	Leveraging edge computing and building lightweight AI models can enable efficient onboard processing and real-time analytics.
Autonomy vs. control	Balancing the autonomy of AI systems and the requirement for human oversight in critical decisions brings about a complex challenge.	Implementing AI systems with adjustable autonomy levels and guaranteeing transparent AI decision-making processes can help maintain the human control.
System integration	Incorporating AI and ML technologies into current spacecraft systems and workflows requires overcoming technical and operational barriers.	Adopting modular design principles and promoting interoperability standards can facilitate the seamless integration of AI technologies into spacecraft systems.
Interoperability	The seamless integration and communication between DTs and various systems or platforms.	Securing robust and reliable technology through rigorous testing and validation. Also, implementing standards and certifications.

But despite these challenges, ongoing research aims to enhance fidelity, efficiency, and security. Advances in cloud computing, edge computing, and AI algorithms aid in addressing computational and data management challenges.

Efforts are also being made to improve the security frameworks protecting these digital systems. Because space missions are complex and have high risks, using DTs in spacecraft health management can help prevent maintenance issues, increase safety, and enhance operational efficiency. As research progresses and addresses these challenges, it is my opinion that integrating DTs into space missions will probably become a standard practice.

14.5.
A Roadmap for Technology Development

Significant milestones have played a role in marking the course toward unprecedented autonomy, efficiency, and reliability. The overarching goal is to create spacecraft that are self-sufficient and capable of long-duration missions with minimal Earth-based intervention. The roadmap depicted in **Figure 14.1** technology outlines a series of phases, each with its own set of milestones and objectives.

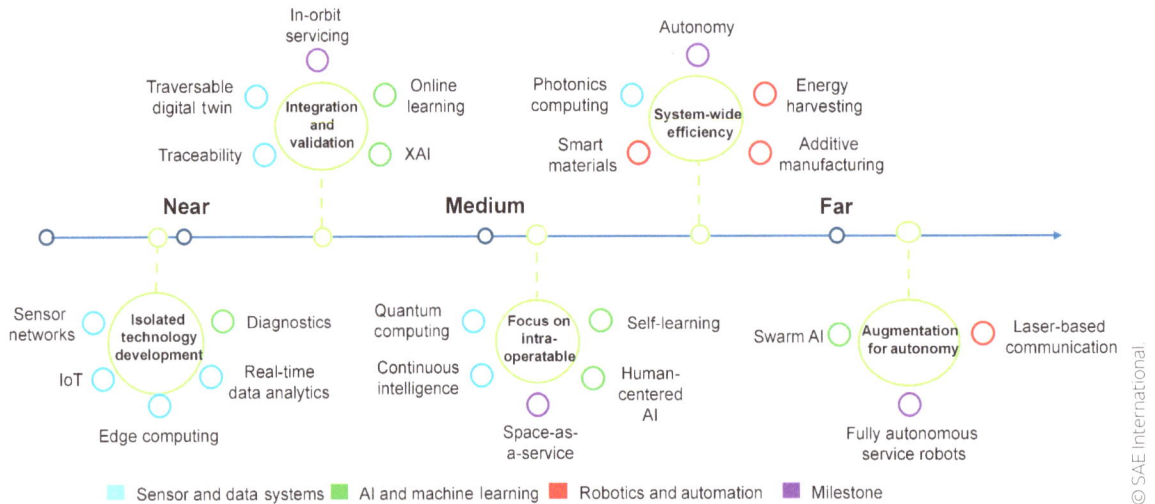

In the initial phase, the focus is on establishing a foundation for integrating technologies like advanced sensor networks, IoT, and edge computing. These technologies enable real-time monitoring, diagnostics, and the ability to respond to system health data. To develop the fundamental abilities required for system integration, this foundation is crucial when conducting any form of predictive maintenance. As we move forward toward reaching more sophistication with in-orbit services, DTs and XAI play a much larger role. These technologies offer the potential to enhance a spacecraft's ability to self-diagnose and repair, extending mission durations and reducing the need for costly and complex rescue or maintenance missions. During this phase, an operational DT acts as a virtual mirror of the spacecraft, enabling various simulations, analyses, real-time decision-making, and future design improvements.

The route toward autonomous capability is further developed in the subsequent phase, where quantum computing, continuous intelligence, and advanced energy solutions become part of the solution. This signifies a critical milestone in achieving a level of autonomy that allows spacecraft to make complex decisions independently, supported by AI systems capable of continuous learning. The role of quantum computing in addressing complex navigational and environmental challenges, coupled with sustainable energy harvesting technologies, underscores the shift toward fully autonomous operations. The phase focused on establishing complete operational independence and achieving the full realization of sustainable autonomy. Here, additive manufacturing for in-space maintenance and laser-based communication networks are significant innovations. The ability to manufacture parts on demand revolutionizes spacecraft maintenance, significantly improving self-sufficiency and mission resilience. Meanwhile, laser-based communication guarantees high-speed data transmission, vital for the management and operation of the spacecraft.

The last phase of the roadmap envisions the expansion of these technologies for deep space exploration. This includes integrating human-centered AI and developing self-sustaining ecosystems.

These milestones are not just technological achievements; they represent the result of decades of innovation. However, many of these opportunities, particularly those revolving around "integration and validation," hinge on integrating technological infrastructure and focusing on resilience. Such integration is difficult and will call for a combination of the right technologies, domain expertise, and partnerships in the New Space initiative. Dealing with data heterogeneity, redundancy, interoperability, and evaluation poses significant challenges.

Most of these technical challenges will relate to data availability and access, making sure they accurately represent reality. For example, for autonomous capabilities to be effective, any delay in data retrieval and processing can hinder timely actions, diminishing any benefits. Therefore, solutions must focus on seamless connectivity, establishing data origins are clear, and creating a reliable source for the DT. Data intelligence should translate raw data into useful information for decision-making. Multimodel simulation can facilitate the debugging of large-scale business scenarios without altering the physical device. Human–machine interfacing will improve interaction efficiency, while quantum will help manage the computational demands and integration of complex system data.

Each milestone on this roadmap enhances individual spacecraft and advances collective knowledge and organization for future space

missions. Implementing autonomous maintenance technologies requires a lot of industry testing and acceptance of associated risks and costs. Reducing cultural effects through standards and collaboration across organizations can clarify operational benefits and promote unified terminology. Designing a specific health management solution around autonomy and maintenance, supported by cross-organizational collaboration, could further aid this endeavor. As health management develops in this field, it promises to make space travel safer, more efficient, and more accessible, opening up new possibilities for scientific discovery, commercial ventures, and exploration.

14.6.
Concluding Remarks

The trajectory of spacecraft system health management is shifting toward greater integration of AI and autonomy. This has profound implications for both spacecraft design and mission planning.

- Integrating advanced health management systems will cause significant improvements in mission safety and reliability. By enabling early detection and mitigation of potential failures, these systems will elevate the overall success rate of space missions.

- DTs and autonomous health management capabilities will bring about a shift in spacecraft design paradigms. Designers must incorporate these technologies from the beginning to enable spacecraft to monitor, diagnose, repair, and adapt to unforeseen challenges.

- Operational efficiency: The application of AI and ML will streamline mission operations, minimizing the requirement for continuous human oversight and intervention. This efficiency gain will allow for more ambitious missions, including deep space exploration and long-duration habitation missions, by ensuring that spacecraft can manage their health effectively.

- As technologies mature, their integration into spacecraft design and operation will develop increased economic viability. This viability will unlock new opportunities for commercial space ventures, including satellite servicing, space tourism, and resource extraction.

- Regulatory and ethical considerations: The advanced autonomy of spacecraft raises important regulatory and ethical considerations. Ensuring the safety and reliability of autonomous systems will require rigorous testing and validation, as well as international cooperation to establish standards and regulations.

Reflecting on the contents of this book, we are in the midst of change. Some of the technological advancements offer a significant improvement in the safety, efficiency, and longevity of systems. This realization encourages us to envision a collaborative relationship between technology and human expertise, rather than technology replacing human roles. The key to this evolution is integrating these advanced concepts with the deep insights and critical thinking unique to humans. Through integrating new technology with human intelligence, we can embark on a new era of missions that are safer, more reliable, and truly innovative.

I will wrap up this book with a thought that reflects the expanding scope of the space industry.

Spacecraft system health management is important not just for scientists and engineers, but for the success of all future space missions. As space exploration advances, the goal is not just to go farther, but to develop smarter technologies that ensure spacecraft reliability and safety. To fully realize the potential of New Space advancements, we must continue to innovate, collaborate across industries, and apply interdisciplinary technology concepts to create more resilient systems. By doing so, we can transform the future of space exploration, ensuring long-term mission success and bringing the benefits of space technology back to Earth.

Index

About the Author

Dr. Samir Khan

Dr. Samir Khan is an Associate Professor at the University of Tokyo, where he leads research in AI-driven control, perception, and planning for autonomous systems. His expertise lies in leveraging machine learning for prognostics

and system health management, with applications across aerospace and other high-reliability sectors. Dr. Khan received his PhD in Control Engineering from Loughborough University, UK, in 2010, where his research laid the groundwork for his later contributions to advanced fault detection and diagnostics.

Dr. Khan's early work at Cranfield University, particularly through the "No Fault Found" project, has had a lasting impact on the aerospace industry, addressing critical challenges in fault detection and system reliability. In 2016, he was awarded the prestigious Japan Society for the Promotion of Science (JSPS) Fellowship, where he developed innovative informatics platforms for next-generation health monitoring services. Currently, his research at the University of Tokyo spans active monitoring, advanced sensor networks, dynamic modeling, and the integration of machine learning for autonomous system management.

Dr. Khan has a strong record of collaboration with leading industrial organizations, including Rolls-Royce, Sumitomo Electric Industries, JAXA, Jaguar Land Rover, BAE Systems, and the UK Ministry of Defence. His work bridges the gap between academic research and practical industry applications, reflecting his commitment to addressing real-world challenges through rigorous research methodologies. His contributions to these fields are reflected in over 50 peer-reviewed publications and numerous invitations to present his work at leading international conferences.

In addition to his research, Dr. Khan is dedicated to educating the next generation of engineers. He has developed and taught courses on digital and analog electronic design, instrumentation, control systems, artificial intelligence, and robotics. His teaching is informed by both his extensive research experience and his industrial collaborations, providing students with a comprehensive understanding of both theoretical concepts and practical applications.

Dr. Khan is a Chartered Engineer, a Fellow of the UK Higher Education Academy, and a member of the Institution of Engineering and Technology (IET) and the Institute of Electrical and Electronics Engineers (IEEE). His academic leadership and international collaborations continue to make significant contributions to the fields of intelligent systems and health management in autonomous technologies.

www.ingramcontent.com/pod-product-compliance
Lightning Source LLC
Chambersburg PA
CBHW050900210326
41597CB00002B/34

* 9 7 8 1 4 6 8 6 0 7 7 9 6 *